William M. Lewis, Jr.

Zooplankton Community Analysis

Studies on a Tropical System

With 36 Illustrations

Springer-Verlag
New York   Heidelberg   Berlin

William M. Lewis, Jr.

Department of Environmental,
Population and Organismic Biology
University of Colorado
Boulder, Colorado 80309
USA

Library of Congress Cataloging in Publication Data

Lewis, William M 1945–
    Zooplankton community analysis: studies on a
tropical system.

    Bibliography: p.
    1.   Freshwater zooplankton—Philippine Islands—
Sultan Alonto, Lake.   2.   Plankton populations—
Philippine Islands—Sultan Alonto, Lake.   I.   Title.
QL143.L48   592   79-17162

The use of general descriptive names, trade names, trademarks, etc.
in this publication, even if the former are not especially identified,
is not to be taken as a sign that such names, as understood by the Trade
Marks and Merchandise Marks Act, may accordingly be used freely by
anyone.

Printed in the United States of America.

9 8 7 6 5 4 3 2 1

ISBN 0-387-90434-4   Springer-Verlag New York
ISBN 3-540-90434-4   Springer-Verlag Berlin Heidelberg

To David G. Frey (Mr. Cladocera)

Preface

This book is based on the premise that the study of ecological communities should be a composite analysis of system properties (community structure, community energetics) and population properties (life history patterns, adaptive strategies) backed by a thorough understanding of the physical–chemical environment. Too frequently community ecology takes a much narrower focus. This may partly be the result of perceived antagonisms between schools of thought in ecology. Despite their rather separate origins, the multiple theoretical and methodological tools that now exist must be applied synthetically to real communities if the progress of the past two decades is to continue into the next two.

This book has a case history format, which increases the opportunity for detailed analysis, although I have attempted to maintain the general perspective of a community ecologist and to draw extensively from the literature whenever it seems profitable to do so. The case history data are for Lake Lanao, a large tropical lake. The main zooplankton data base used in the analysis is entirely original and unpublished, although the detailed supporting data on the physical–chemical environment and the phytoplankton community have been presented in numerous journal articles and are thus abstracted or used selectively to meet the needs of zooplankton community analysis.

Since the case history is a tropical one, the elementary data themselves may be of greater interest than they might otherwise be. I have tried to anticipate the probable interest of readers in temperate–tropical comparisons and to draw on whatever additional meager information is available on tropical freshwater zooplankton communities, including my own work on tropical lakes other than Lanao.

The community analysis relies heavily on an hypothesis-testing approach, with attendant use of simple and multivariate statistical procedures. I have often chosen the simplest possible statistical approaches, however, in the belief that complex procedures frequently offer little additional information but cause great difficulty for the reader in making an independent rational evaluation of the data. I have also placed considerable emphasis on the analytical value of some simple derived variables, such as rates of change, that I believe to be of great use in community analysis.

I am indebted to a number of persons and institutions for support or assistance in the preparation of this work. Professor D. G. Frey sponsored and supported my initial plankton community studies and first directed my attention to tropical plankton communities. I owe a great deal to the National Science Foundation for its support of my plankton community work over an extended period on several different lakes. Numerous competent and dedicated individuals have worked with me and thus have contributed to the completion of this work. I am most indebted to Mr. Rodrigo Calva, Mr. James T. Hunter, and Miss Lili Silva in this regard. I have on several occasions called upon the taxonomic expertise of Professors B. Pejler, B. Bērziņš, F. Kiefer, U. Einsle, V. Kořínek, and Mr. J. F. Saunders, who generously provided me with their expert opinions. I spent a short but extremely profitable time with Dr. D. W. Schindler at the very beginning of my work which has unquestionably affected my outlook and methodology ever since. My wife Cornelia has on many occasions helped me without credit or pay, for which I am grateful. I must also mention Professor M. C. Grant, who has provided me with valuable intellectual stimulation and a broad range of consultation services ranging from statistics to ecological theory, and Dr. Robert Epp for his insights into metabolic characteristics of invertebrates. For my Philippine field work, on which so much of this book is based, I am indebted to Mindanao State University, and particularly to Mila Medale of the Biology Department and President Mayaug Tamano for use of facilities and to the Manila Office of the Ford Foundation for valuable logistical support. Miss Lynn Weatherwax has made the struggle of manuscript preparation as easy as it ever can be, for which I am grateful. All of the figures were expertly prepared by Mr. J. R. Tolen. The University of Colorado provided essential supplementary computer funds which have greatly facilitated my work.

Boulder, Colorado   William M. Lewis, Jr.
August 1979

Contents

Chapter 1

Introduction

Lindeman's (1942) presentation of the trophic dynamic concept was followed, after some initial resistance (Cook, 1977), by a surge of interest in energy flow as a unifying concept in ecology. Studies inspired by the trophic dynamic concept were often broad in scope and more likely to demonstrate trends than mechanisms. More exacting studies of single species obtained better resolution but sacrificed the community perspective inherent in Lindeman's original work. Subsequently, new theoretical work dealing with community structure and the adaptive strategies of species populations to a large extent undermined the specific focus of ecology on energetics. These events are nicely summarized by Hutchinson (1978).

Freshwater ecology has not been precisely in phase with other branches of ecology in these trends. General interest in freshwater community structure, for example, lagged about 10 years behind the theoretical advances in this field. Also there was in the late 1950s and 1960s a sudden rejuvenation of interest in energy flow at the first trophic level following the technical innovations associated with the use of carbon-14. There is no question, however, that interest in energy flow, particularly in herbivores and higher trophic levels, has now to a large extent been displaced by interest in the structure and adaptive patterns of freshwater communities. Studies of community structure and adaptive strategy have successfully drawn attention away from energy flow partly because they have often proven to be a rich source of mechanistic detail which is especially satisfying to those who wish to understand the interspecific interfaces which shape communities.

As studies of community structure progress, however, we begin to see new requirements for detailed information on energy flow through the higher trophic levels (e.g., see Rigler and Cooley, 1974). Information on energy flow is extremely useful for a complete evaluation of community structure and species adaptation, as energy provides a powerful means for comparison of species, both within and between trophic levels. Moreover, such phenomena as predation and competition, which determine community structure, are in large part evolutionary exercises in energy allocation as accomplished by adaptation.

A great potential exists for composite studies of community structure, energy flow, and adaptation in aquatic communities and especially in the plankton, where the problems of measuring the relevant variables are most tractable. The data requirements are very great but certainly not impossible. Unlike studies more explicitly concerned with energy flow, such composite studies obviously cannot focus on one or two important species on the basis that these species account for the bulk of energy flow through a given trophic level. On the contrary, composite studies must deal with the entire species complex in order to demonstrate why certain demographic strategies, morphologies, and adaptive patterns are successful in accounting for large portions of production or biomass while others are not. Similarly, composite studies are more demanding of data than are studies more explicitly concerned with community structure or adaptation, as the analysis of energy flow requires a firm grasp of dynamics. Nevertheless, there is much overlap in the data requirements for the different approaches, and the additional effort required to obtain a data base that will support a composite approach is likely to be repaid by much additional insight.

The present work attempts to combine the strengths of community analysis from the perspectives of structure, energy flow, and adaptive strategy. The analysis deals specifically with Lake Lanao (Philippines), but the findings are put into a more general context whenever possible. For several reasons, Lake Lanao is especially advantageous for the purpose at hand. Physical and chemical factors are not so tightly coupled to biological processes as they might be in a temperate lake (Lewis, 1974), and thus the statistical exploration of mechanisms is much more feasible than it would be for a temperate lake. In addition, the plankton community is relatively simple, so that the demographic analysis of individual species is tractable. The large size of the lake is also favorable because it ensures the existence of a true pelagic plankton community that will not be easily confused with the adjacent littoral community, yet the lake is not so large that the plankton of open water cannot be dealt with as a unit. Finally, continuously favorable weather allows steady data collection and ensures high biological activity as well, both of which facilitate the development of a detailed and revealing data base.

Chapter 2

General Description of Lake Lanao

Lake Lanao is located in the southern Philippines on the Island of Mindanao (8°N, 124°E) (Fig. 2–1). The lake was formed by lava blockage of a tectonically formed basin (Fig. 2–2). Although good estimates are not yet available, Lanao is apparently very old, perhaps dating back as far as the Tertiary (D. G. Frey, personal communication). The lake contains a swarm of 20 or more endemic cyprinid species (Myers, 1960; Kosswig and Villwock, 1964), but there appears to be only limited endemism among other elements of the fauna.

Physical characteristics of Lake Lanao are summarized by Frey (1969). The lake has a maximum depth of 112 m, a mean depth of 60.3 m, an area of 357 km², and a replacement time of 6.5 years. The elevation of the lake surface is 702 m asl. The shores are for the most part steeply sloped, which greatly restricts the extent of the littoral zone. The littoral zooplankton, which differs somewhat in composition and dynamics from the pelagic zooplankton, will not be considered here.

The watershed of Lake Lanao is sparsely populated and the rivers flowing into the lake carried very low nutrient loads when the lake was studied in 1970–71. Nutrient budgets and water chemistry have therefore probably changed little over the last several thousand years. Changes will unquestionably occur in the future as timber is cut and population density increases.

Lanao is a warm monomictic lake (Lewis, 1973a). An annual circulation period occurs during January and February, when the entire water column is approximately 24°C. The seasonal cooling is caused by reduced daily insolation resulting from heavy cloud cover at this time of year. The lake is stratified the rest of the year, with hypolimnetic temperatures at or just above

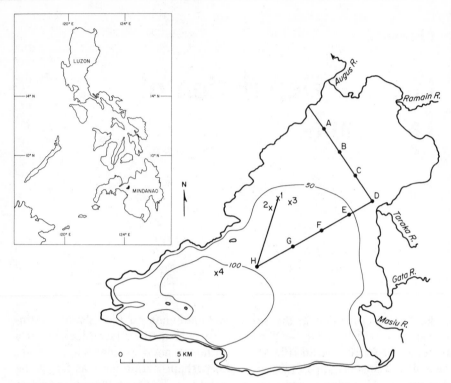

Figure 2–1. Map showing the sampling stations and 50- and 100-m contour lines of Lake Lanao.

Figure 2–2. Lava dam which blocks the south end of the Lake Lanao basin. Foreground shows gillnet suspension device of Muslim fishermen.

24°C and epilimnetic temperatures about 2°C higher except very near the surface where temperatures may reach 29°C or more in calm weather.

Important nonseasonal variations in the depth of mixing occur during the stratification period in Lake Lanao (Lewis, 1973a). The epilimnion of the lake is very thick (40–60 m), mainly because the maximum possible density difference between layers is lower than it would be in a typical temperate situation, so a given wind strength will drive the thermocline deeper. The thickness of the epilimnion is established during windy weather near the beginning of the stratification season. Later, during extended periods of moderate to light winds and high insolation, there is an accumulation of heat in the uppermost portion of the epilimnion. Since the wind energy at such times may be insufficient to mix this heat to the full depth of the epilimnion, a relatively stable secondary thermocline is likely to form within the epilimnion. In fact it is not uncommon for two such secondary thermoclines to form, one above the other. I have argued that this phenomenon will prove to be common in tropical lakes deep enough to stratify stably (Lewis, 1973a), and recent work on Lake Valencia in Venezuela tends to support this idea (Lewis, in preparation), as does the work of Richerson et al. (1975) on Lake Titicaca.

When a thick epilimnion is split by a thermocline, the upper and lower portions diverge chemically. The lower portion serves as a nutrient trap, while primary production is limited to the upper zone. The secondary thermocline persists until windy weather comes, which for Lake Lanao may be from 1 to several weeks (Fig. 2–3). When the entire epilimnion is homogenized again, nutrients are returned to the surface from the compartment below the secondary thermocline, stimulating primary production. I have called this nonseasonal remixing process "atelomixis" (Lewis, 1973a) and I have hypothesized that it contributes in a major way to the high primary production of Lake Lanao, which is not particularly rich in nutrients. If mixing to the primary thermocline occurred continuously, phytoplankton would suffer light deprivation, and if mixing to the shallower secondary thermocline occurred continuously, phytoplankton would suffer marked nutrient depletion. Alternation of deep and shallow mixing provides a more optimal average resource environment and thus boosts primary production.

Primary production in Lake Lanao is limited by sunlight during periods of deep mixing, especially during complete circulation, and at other times by nutrients, specifically inorganic nitrogen (Lewis, 1974). The overall primary production of the lake is very high (annual mean, 1.7 gC/m² · day net, 2.6 gC/m² · day gross). Inorganic nitrogen is frequently undetectable (< 1 μg/liter) in the upper water column, whereas molybdate-reactive phosphorus is always above 10 μg/liter. The solids content and major ionic ratios are unexceptional (105 μmho/cm at 25°C; methyl orange alkalinity, 51 mg/liter). Average 1% light level occurs at 12 m.

There are two fundamental physical–chemical contrasts between Lake

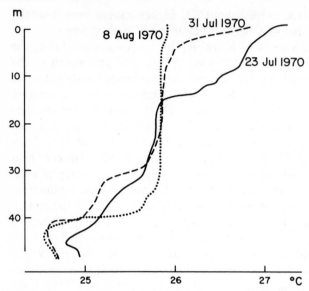

Figure 2–3. Multiple thermoclines and thermocline displacement during the stratification season in Lake Lanao. Storms on 24 July and 6 August displaced the higher thermocline and then fused it with the lower one.

Lanao and most temperate lakes. First, the average amount of incident light and average temperature of the water column are both higher in Lake Lanao and other lowland tropical lakes than in temperate lakes. Constantly high incident light provides sufficient energy for extremely rapid photosynthesis whenever the water column is not circulating so deeply as to remove phytoplankton from the lighted zone, and provided that nutrients are available in sufficient amounts. Constantly high temperatures prevent thermal depression of metabolic activity either seasonally or vertically in the water column. It is important but still difficult to bear in mind for those whose experience with lakes is entirely temperate that the temperature below the thermocline in a lowland tropical lake is extremely high (24°C in Lake Lanao). Temperature in the tropical plankton environment appears to be almost negligible as a source of biotic variation. This is highly desirable from the viewpoint of ecological analysis and also has considerable implications for the adaptive strategies of species and the rates of biologically mediated processes such as decomposition.

The second physical–chemical contrast between lakes such as Lanao and most temperate lakes has to do with variation in biologically critical variables. The question of variation is most easily considered in two parts: (1) seasonal variation and (2) nonseasonal (aperiodic) variation. In general it can be shown that tropical lakes have seasonal patterns and that the biological effects of this seasonality are remarkably similar to those one sees in temperate lakes. In contrast, surprisingly critical differences exist between Lake Lanao and most temperate lakes in the nature and biological effects of non-seasonal variation.

Temperate lakes show a very pronounced annual cycle in essentially all variables because of the drastic seasonal weather changes. It is now becoming obvious, however, that most tropical lakes show a very similar tendency toward annual patterns. Although we do not yet have enough information to make definitive statements, it seems increasingly probable that seasonality associated with an annual circulation period will be suppressed only in tropical lakes which are not deep enough to stratify (i.e., less than about 15 m for small lakes). The widely used scheme of lake type distributions proposed by Hutchinson and Loffler (1956) prior to the availability of much comprehensive tropical lake data is therefore misleading because of its emphasis on the existence of significant numbers of lakes of the so-called oligomictic type, which stratify but lack an annual circulation pattern because they circulate at irregular intervals longer than 1 year. Such lakes appear to be quite rare. Even in lakes which are not deep enough to stratify, other factors such as seasonal changes in flushing rate will often impose an annual pattern (e.g., Lake Chad; Iltis and Compere, 1974).

Although temperature does not vary much in Lake Lanao, seasonal temperature changes are sufficient to cause an annual breakdown in thermal structure followed by complete mixing. Complete mixing deprives phytoplankton in the water column of light and thus suppresses growth. This in turn affects the higher trophic levels. Until some thermal stability is restored, there is a seasonal depression of primary production, phytoplankton biomass, and herbivores. In this sense the annual cycle of Lake Lanao and probably many other tropical lakes is surprisingly similar to the familiar temperate cycle.

Nonseasonal or aperiodic changes are imposed on seasonal cycles in all lakes. Because of its strong tendency to develop temporary secondary thermoclines, Lake Lanao shows a large amount of nonseasonal variation under the influence of temporary and irregular weather changes. This is not as true of temperate lakes, in which the density barrier to mixing seldom changes radically once it is established during the warm season. The phenomenon which is perhaps biologically most comparable in temperate lakes to the radical nonseasonal changes in depth of mixing in Lake Lanao is the occasional escape of nutrient-rich water from the hypolimnion into the epilimnion when an internal wave of oscillation brings the hypolimnion to the surface. Such an event is either so unusual or so temporary in its effect that it does not result in the frequent major changes of nutrient availability and phytoplankton growth that one sees in Lake Lanao as a result of secondary thermocline formation. A more or less steady decline in nutrient availability during stratification in temperate lakes contrasts with the numerous irregular sequences of nutrient enrichment and nutrient depletion in Lanao. These frequent changes in the tropical lake are extremely important in reducing environmental predictability, in interrupting phytoplankton succession, and in increasing the overall potential of the lake for synthesizing biomass at the first trophic level by promoting more effective nutrient recycling. Counterintuitively to our general notions of the tropics, critical physical–chemical conditions for the primary producers in Lake Lanao are more irregular during the growth season than in most temperate lakes that have been studied.

Chapter 3

The Phytoplankton Community

Community Composition

The phytoplankton community of Lake Lanao is composed of 70 euplank-tonic species, including Cyanophyta (12), Euglenophyta (4), Chlorophyta (44), Chrysophyceae (1), Bacillariophyceae (4), Dinophyceae (3), and Cryptophyceae (2) (Fig. 3–1). Composition similar to this appears to be very widespread in tropical lakes that are not highly saline or extremely shallow (Lewis, 1978a). The data that are currently available indicate that tropical lakes in general contain between 50 and 100 euplanktonic autotroph species which reach detectable abundances (greater than 1/ml) at least once per year. In Lake Lanao, approximately half of the species can be found in an integrated sample of 1 ml taken at any time of the year.

Neither Lake Lanao nor tropical lakes in general show any evidence of being richer in phytoplankton species than lakes of the temperate zone (Lewis, 1978a). In fact the available evidence indicates that the reverse is true, as the Chrysophyceae, which are almost absent in tropical plankton, become an increasingly important component of phytoplankton communities at high latitude and enrich the species composition. Certainly phytoplankton communities show no evidence of the increased species richness at low latitudes which is characteristic of some other community types. The multiple hypotheses commonly used to account for latitudinal gradients in lizard, bird, and tree diversity (summarized by Pianka, 1978) are difficult to evaluate for phytoplankton communities but might easily be considered to provide a priori grounds for expecting latitudinal gradients in phytoplankton species richness which are not in fact observed. Hypotheses based on equi-

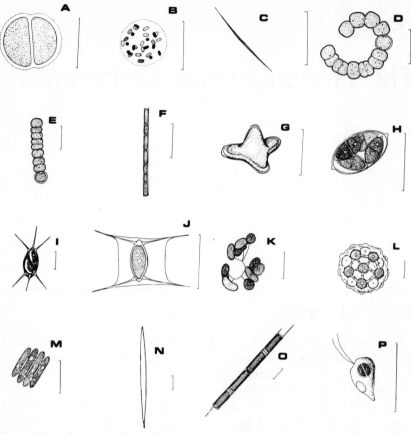

Figure 3–1.   Common phytoplankton species in Lake Lanao. (A) *Chroococcus minutus;* (B) *Aphanothece nidulans;* (C) *Dactylococcopsis fascicularis;* (D) *Anabaena spiroides;* (E) *A. sphaerica;* (F) *Lyngbya limnetica;* (G) *Tetraedron minimum;* (H) *Oocystis submarina;* (I, J) *Chodatella subsalsa* (showing extremes of seasonal variation); (K) *Dimorphococcus lunatus;* (L) *Coelastrum cambricum;* (M) *Scenedesmus* sp.; (N) *Nitzschia baccata;* (O) *Melosira granulata;* (P) *Rhodomonas minuta.* Scale marks are 10 μm long.

librium conditions may be irrelevant to phytoplankton communities, however, because of the rapid pace of physical–chemical changes in plankton communities at all latitudes. The general relevance of nonequilibrium conditions stressed by Connell (1978) is likely to be especially great for phytoplankton communities and will probably be the basis for satisfying explanations of community species richness.

Most of the species which are important in Lake Lanao are very widely distributed. The overlap at the generic level between tropical lakes, even on different continents, is extremely high (Table 3–1). Overlap in species composition is also very high but is more difficult to quantify because of the un-

Table 3–1.   Phytoplankton Community Comparisons between Lake Lanao and Other Lakes[a]

	Mean	Standard Deviation
Number of genera		
Lake Lanao	42	—
10 Tropical lakes, mean[b]	21	5
10 Temperate lakes, mean	36	16
Generic overlap (%)		
Lanao with 10 tropical lakes, mean	79	11
Lanao with 10 temperate lakes, mean	45	5
Biomass (mg/m², wet)		
Lake Lanao	24,000	—
11 Tropical lakes, mean	14,000	13,500
11 Temperate lakes, mean	7,500	3,810

[a] Summarized from Lewis (1978a).
[b] Probably somewhat underestimated because of small number of samples from each lake.

certainty of species assignments in some taxa. In fact the overlap is suffi-ciently high that it seems justifiable to characterize most tropical phytoplankton communities as part of a pantropical phytoplankton assem-blage that does not appear to be affected significantly by geographic barriers.

In addition to the large taxonomic overlap between tropical phytoplankton communities in different lakes, there is also a surprising amount of overlap between tropical and temperate phytoplankton communities (Table 3–1). The overlap is less, however, than the overlap within the tropics. Although there may be much hidden biochemical diversity in these taxa, the phyto-plankton do not show much evidence of isolation and evolutionary diver-gence.

Biomass

Because the mixed layer of Lake Lanao is generally quite thick, even when a secondary thermocline is present, growing phytoplankton biomass is effi-ciently dispersed through the upper portion of the water column. Thus even though the water is relatively transparent, summation of phytoplankton bio-mass under a unit area shows a very high standing crop. As indicated in Table 3–1, the algal standing crop of Lake Lanao may be above the average for tropical lakes (polluted, shallow, or saline lakes are excluded from consideration here). The average for tropical lakes reported in Table 3–1 is probably low, however, because one group of these lakes was sampled only during the circulation period when biomass was likely to have been considerably below the annual average (Lewis, 1978a). The Lake Lanao figure is an annual average. The averages in Table 3–1 are, of course, intended to be indicative rather than definitive, especially in view of the wide scatter within latitudinal zones, as indicated by the standard deviations.

Tropical lakes in general may have a substantially higher average standing stock of phytoplankton than temperate lakes, although the ranges definitely overlap. Lower standing stock in temperate lakes is partly accounted for by severe winter depression of phytoplankton biomass in the temperate zone, but average phytoplankton biomass also appears to be lower in temperate lakes even when when averages are computed from the ice-free season only (Lewis, 1978a).

Even though the phytoplankton biomass of Lake Lanao is divided among seven major taxa, three of these account for most of the total biomass (Cyanophyta, 19%; Chlorophyta, 35%; Bacillariophyceae, 37%). Dominance of these three taxa is typical of tropical lakes which are not saline or extremely shallow (Lewis, 1978a). Very shallow and saline lakes tend to be dominated more exclusively by the Cyanophyta.

Succession

Succession of the phytoplankton community in Lake Lanao occurs in episodes of variable length (Lewis 1978b). An episode is initiated when the upper portion of the water column is enriched by the breakdown of temporary stratification, by the lowering of the primary thermocline, or by seasonal mixing. A nutrient pulse (10–50 μg/liter NO_3–N) resulting from one of these events leads to dominance by diatoms and cryptomonads as soon as deep mixing is reduced enough to allow the average cell to stay in the lighted zone long enough for significant net photosynthesis. As nutrients are depleted and turbulence is reduced during calm weather, diatoms and cryptomonads tend to be replaced by green algae. The greens are replaced by blue-green algae and finally by dinoflagellates as the trend continues toward the extremes of nutrient depletion (< 1 μg/liter NO_3–N) and minimal turbulence (Fig. 3–2). The sequence is roughly the same as that which is thought to occur in temperate lakes (see Reynolds, 1976). The sequence may be interrupted or various phases of it may be squeezed or lengthened according to the specific conditions of a particular episode. In addition, individual species do not show perfect fidelity to their division or class (Lewis, 1978b).

Underlying the successional sequence are relationships between the surface:volume ratio of biomass units for individual taxa and the position of these taxa on gradients of nutrient availability and turbulence (Lewis, 1978b). In early succession, when sinking rates are low and nutrient availa-

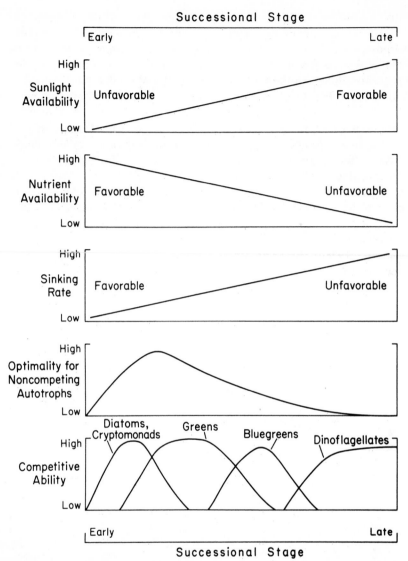

Figure 3–2. Diagrammatic representation of the phytoplankton succession sequence and associated physical–chemical changes in Lake Lanao (slightly modified from Lewis, 1978b).

bility is high, species characterized by low surface:volume ratios are favored. During periods of nutrient depletion and minimal turbulence, species with high surface:volume ratios tend to predominate, presumably because of their superior ability to scavenge scarce nutrients. Although this trend is unquestionably valid for Lake Lanao, it runs counter to the generalizations formulated by Margalef (1967), who suggests that cells with high surface to volume ratio will predominate early in succession because of their more rapid response capability to improved conditions. Preliminary indications from Lake Valencia in Venezuela indicate that it will conform to the Lake Lanao pattern rather than the pattern proposed by Margalef. Additional examples are needed before generalizations on this subject will be very meaningful.

The taxonomic pattern in the successional sequence for Lake Lanao is partly related to a difference in average surface : volume ratio of the biomass units in the different major taxa. The bluegreen algae, for example, have much lower surface : volume ratios than the diatoms.

The major contrast between succession in Lake Lanao and succession in familiar temperate lakes is that the number of successional episodes in a single stratification season is much greater in Lake Lanao. This has been demonstrated quantitatively in a comparison between Lake Lanao and Lake Erken (Lewis, 1978b). In Lake Lanao, significant pulses of nutrients at the surface of the lake set succession back to the early stages more frequently and thus sustain the community in the younger phases of succession for a greater percentage of the growing season. The quantitative comparison with Lake Erken indicated that Lanao was set back to stage one of succession (diatoms) three times as often as Lake Erken, but that the two lakes reached the penultimate stage (bluegreens) equally often. This implies more numerous successional sequences of shorter average duration in Lanao. It follows that frequent successional setbacks lead to extremely complex demographic patterns for individual phytoplankton species in Lake Lanao.

Chapter 4

Description of the Zooplankton Community

Community Complexity

The complete list of zooplankton species is given in Table 4–1 and Figure 4–1 shows the species and developmental stages on a common scale. The number of species is surprisingly small, particularly by comparison with some familiar temperate lakes. For example, Nauwerck (1963) lists 36 rotifer species, 16 cladoceran species, and 12 copepod species for Lake Erken, Sweden. The great majority of these species are quantitatively insignificant, however. In small lakes such as Erken, species lists may be deceptively long due to the greater likelihood of inclusion of littoral species in the list. This and other factors make comparisons difficult.

For the 15 lakes on Java, Sumatra, and Bali visited by the Sunda expedition, Ruttner's (1952) entire species list contains only 23 rotifers, 13 cladocerans, and 6 copepods, despite the considerable taxonomic expertise that was brought to bear on the samples. Mean numbers of rotifer, cladoceran, and copepod species per lake in these 15 lakes were 7.2, 1.6, and 2.0, respectively. This compares closely with the numbers in Lanao (7, 4, and 2). The zooplankton fauna of Lake Mainit, another large Philippine lake, is of similar or even simpler organization (Lewis, 1973b). The Lanao assemblage is thus not unusual among tropical lakes.

Pennak (1957) has compared the mean number of species at an instant in time with the composite number over a year in a number of temperate lakes. In 27 Colorado Lakes, the average numbers of rotifer, cladoceran, and copepod species at an instant in time were 4.8, 1.6, and 1.3, respectively. For other scattered temperate lakes, the means were 5.5, 2.8, and 2.7. These fig-

Table 4–1. List of Species and Developmental Stages with Their Average Weights and Abundances over the Study Period

Species/stage	Mean individual/liter	Mean individual/m² (1000s)	Number counted per week	Mean length (μm)	Mean biomass per individual (μg, wet)	Mean total biomass (μg/liter, wet)
Copepods						
Thermocyclops hyalinus						
Nauplius 1	12.28	552.60	1,842	86	0.08	0.98
2	20.28	912.60	3,042	115	0.21	4.26
3	18.76	844.20	2,814	127	0.25	4.69
4	15.82	711.90	2,373	160	0.50	7.91
5	13.81	621.45	2,072	171	0.55	7.60
6	7.90	355.50	1,185	205	0.81	6.40
Σ	88.87	3,999.15	13,331	—	—	31.84
Copepodid I	7.78	350.10	1,167	320	1.43	11.13
II	5.91	265.95	887	342	1.87	11.05
III	4.82	216.90	723	422	2.84	13.69
IV	4.85	418.25	728	480	3.69	17.90
V	2.59	116.55	389	560	5.30	13.73
Σ	25.97	1,168.55	3,896	—	—	67.50
Adult ♂	6.34	285.30	951	547	3.41	21.62
♀	4.54	204.30	681	592	6.32	28.70
Σ	10.88	489.60	1,632	—	—	50.32
Eggs	10.19	458.55	1,529	57	0.09	0.92
Tropodiaptomus gigantoviger						
Nauplius 1	0.20	9.00	30	180	0.63	0.13
2	0.47	21.15	71	227	1.19	0.56
3	0.78	35.10	117	260	1.28	1.00
4	0.91	40.95	137	287	1.71	1.56
5	0.59	26.55	89	324	2.63	1.56
6	0.59	26.55	89	357	3.18	1.88
Σ	3.56	160.20	534	—	—	6.69

Copepodid I	0.94	42.30	141	402	4.54	4.27
II	0.76	34.20	114	522	6.41	4.87
III	0.75	33.75	113	635	8.18	6.14
IV	0.78	35.10	117	777	15.0	11.70
V	0.79	35.55	119	914	18.3	14.47
Σ	4.04	181.80	606	—	—	41.45
Adult ♂	1.24	55.80	620	967	21.8	27.03
♀	0.88	39.60	440	1,128	44.0	38.72
Σ	2.13	95.85	1,056	—	—	65.75
Eggs	1.69	76.05	845	138	0.87	1.47
Cladocera						
Diaphanosoma modigliani[a]	4.79	215.55	719	553	4.0	19.16
Eggs/embryos	1.40	63.00	210	215	0.61	0.85
Moina micrura	0.43	19.35	65	534	6.0	2.58
Eggs/embryos	0.19	8.55	29	150	0.69	0.13
Bosmina fatalis	1.50	67.50	225	306	2.4	3.60
Eggs/embryos	0.31	13.95	47	130	0.37	0.11
Rotifera						
Conochiloides dossuarius	17.48	786.60	2,622	126	0.15	2.62
Egg	4.28	192.60	642	49	0.044	0.19
Hexarthra intermedia	4.27	192.15	641	101	0.23	0.98
Egg	1.05	47.25	158	49	0.044	0.05
Polyarthra vulgaris	0.81	36.45	122	77	0.29	0.23
Egg	0.20	9.00	30	49	0.044	0.01
Keratella procurva	0.68	30.60	102	158	0.13	0.09
Egg	0.13	5.85	20	43	0.046	0.01
Keratella cochlearis	3.96	178.20	594	119	0.070	0.28
Egg	0.72	32.40	109	43	0.046	0.03
Trichocerca brachyurum	0.26	11.70	39	146	0.49	0.13
Egg	0.06	2.70	9	50	0.047	0.003

Table 4–1. (continued)

Species/stage	Mean individual/liter	Mean individual/m² (1000s)	Number counted per week	Mean length (μm)	Mean biomass per individual (μg, wet)	Mean total biomass (μg/liter, wet)
Tetramastix opoliensis	4.33	194.85	650	486	0.13	0.56
Egg	0.49	22.05	74	76	0.082	0.04
Diptera						
Chaoborus (Eckstein 1)	0.160	7.2	360	3,528	326.0	52.16
Shrimp	0.00380	0.17	8.5	6,000	1,809.0	6.87
Fish larvae	0.00015	0.007	0.34	5,000	1,520.0	0.23
Total herbivores	174.0	7,828.0	26,843	—	—	306.48
Total carnivores	0.160	7.2	360	—	—	52.39

[a] Includes some *D. sarsi* as well.

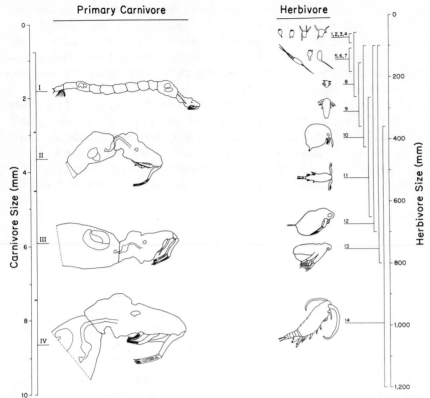

Figure 4–1.   Zooplankton of Lake Lanao, drawn on a common scale. Primary carnivores include the four instars of *Chaoborus*. Herbivores include two *Keratella* species (1,2), *Polyarthra* (3), *Hexarthra* (4), *Tetramastix* (5), *Conochiloides* (6), *Trichocerca* (7), *Thermocyclops* nauplius (8), *Tropodiaptomus* nauplius (9), *Bosmina* (10), *Thermocyclops* copepodid/adult (11), *Diaphanosoma* (12), *Moina* (13), and *Tropodiaptomus* copepodid/adult (14).

ures are very close to the numbers for Lake Lanao and for the Sunda lakes. The species numbers of copepods and Cladocera obtained by Patalas (1971), who examined 45 Ontario lakes for crustaceans at a single time of year, are close to Pennak's figures (mode, 9 species; range 4–14 species of Copepoda plus Cladocera). Patalas found a definite increase in number of species with lake size, and, in another study (Patalas, 1975), documents considerably higher species richness in very large lakes. These large lakes appear to owe their richness in species at least partly to size.

Pennak observed that the composite number of species is at least double the total for a single sample. This increase for samples taken at several times of the year is not typical of Lake Lanao. These comparisons, which are still only approximate, indicate that the number of reasonably abundant species in tropical plankton environments at an instant in time is no greater and may even be less than in comparable lakes at higher latitude, and that the composite annual numbers of species which reach detectable abundance may well be considerably greater in temperature lakes than in tropical ones. This may be attributable to the greater suitability of the temperate lacustrine habitat for temporal separation of species.

The relation of community complexity to latitude is discussed by Green (1972), but Green's analysis clearly includes species which are not euplanktonic. This increases the apparent complexity at any given latitude and introduces other complications related to the size of lake and location of sampling. The average number of euplanktonic species in tropical lakes is probably quite low, as in Lake Lanao, but a satisfactory overview is difficult at present.

Copepods

The first zooplankton samples from Lake Lanao were taken by Woltereck (1941) during the Wallacea Expedition in 1932. The cyclopoids were examined by Kiefer (1938), who found *Thermocyclops hyalinus* in large numbers. *T. hyalinus* is distributed throughout the Asian, American, and African tropics as well as temperate Europe (Kiefer, 1929; Coker, 1943). There is considerable size variation within the species. Females from temperate lakes are as much as 50% longer than females from tropical lakes (Kiefer, 1938). Einsle (1970) has shown that females from European populations of *T. hyalinus* average 821 μm in length (tip of first segment to tip of urosome, discounting caudal rami), whereas females of six African populations average 668 μm. The females from Lake Lanao average only 597 μm (Table 4–1).

In addition to *T. hyalinus,* Kiefer also found a small number of specimens of a slightly smaller but very similar species, which he named *T. wolterecki.* *T. wolterecki* is presumably endemic to Lanao, although it has not been studied since Kiefer's description. The adult females in Kiefer's samples averaged 10–20% shorter than *T. hyalinus.* The two species differ morphologically only with regard to length ratios in the furcal spines and the absence of the minute teeth on the connecting plate of leg 4. *T. wolterecki* thus appears to be a sibling species of *T. hyalinus* or possibly even just a morphotype. The presence of a copepod species of very limited distribution together with one of cosmopolitan distribution is not uncommon (Kiefer, 1938; Hutchinson, 1967), although the ecological reasons for this remain unclear.

Dr. U. Einsle kindly compared a mixed sample taken in 1970 from Lake Lanao with Kiefer's original material. He found only *T. hyalinus* in the 1970 sample but confirmed the presence of *T. wolterecki* in Kiefer's original sample, and is of the opinion that it is a valid species. Samples from various times of the 1970–71 period were examined but *T. wolterecki* was consistently rare or undetectable. The maximum percentage of *T. wolterecki* among adult females occurred in January, when about 20% seemed to be *T. wolterecki*. The present study does not distinguish between the two species because *T. wolterecki* apparently constitutes less than 5% of the total and is so close to *T. hyalinus* that dissection is required for identification.

Lanao also contains the calanoid *Tropodiaptomus gigantoviger,* which was described by Brehm (1933) from samples taken during the Wallacea Expedition and is evidently endemic to Lanao. Since calanoids tend to be very localized in distribution (Hutchinson, 1967), the presence of an endemic calanoid is not particularly remarkable. Professor F. Kiefer personally confirmed that the species collected in 1970–71 was in fact the same as that collected by the Expedition in 1932.

Although the vast majority of tropical lakes contain at least one cyclopoid species, the calanoids are much more irregular in distribution and are missing in many large lakes that appear to provide suitable conditions. Even on Mindanao, the large Lake Mainit, which is similar to Lake Lanao in general water quality and plankton composition, lacks a calanoid (Lewis, 1973b). Of the 15 Sunda lakes visited by Ruttner (1952), all but Lake Toba lack calanoids. The presence of a calanoid thus distinguishes the zooplankton of Lake Lanao from many other lakes where these large planktonic grazers are absent.

Rotifers and Cladocera

Woltereck (1941) reports only *Conochiloides, Brachionus forficulata,* and *Keratella valga* in the Lanao samples taken by the Wallacea Expedition. Curiously, *B. forficulata* and *K. valga* never turned up in the 1970–71 samples. The rotifers which did appear in 1970–71 were identified by B. Bērziņš and B. Pejler and are listed in Table 4–1. All of these species are widely distributed.

The midlake samples taken by the Wallacea Expedition contained *Diaphanosoma modigliani,* which was also present in the samples from 1970–71 according to taxonomic work done on my samples by Dr. V. Kořínek. Kořínek also found *D. sarsi,* but the two *Diaphanosoma* species are not distinguished in the demographic studies. *D. sarsi* was not reported by Woltereck (1941), but this could well have been an oversight due to the similarity of the species. A *Moina* species was also present in the samples from 1932 and 1970–71. Kořínek identifies this as *M. micrura.* Woltereck also found this species but lists it as *M. makrophthalma,* a synonym (Goulden, 1968). Woltereck found *Bosminopsis dietersi* and *Bosmina longirostris* in the pelagic zone, but Kořínek found only *Bosmina fatalis* in the 1970–71 samples, and this is almost certainly identical with Woltereck's *B. longirostris*. The absence of *Bosminopsis* in the 1970–71 samples could possibly be due to a change in species composition since 1932.

Miscellaneous Zooplankton

Lanao also contains a large population of *Chaoborus* comprised of only one species which is, according to an identification made for me by Dr. J. Stahl, identical to the species which was partly described by Eckstein (1936) but never named and which I will refer to as Eckstein Form 1 (Lewis, 1975).

The large plankton trap that was used (45 liters) also captured small fish fry and shrimp. The fish fry belong to several species including the native cyprinids and the introduced goby (*Glossobius giurus*), but are not distinguished by species here. The prawn *Caridina nilotica* is very common in the littoral zone and is probably the species which appears most often in the plankton trap at midlake. F. Chace (personal communication) has also identified the large palaemonid *Macrobrachium latidactylus* from nearshore samples taken by the *Albatross* Philippine Expedition (1907–1910), however.

Chapter 5

Methods

Sampling

Poor sampling design has been perhaps the most serious defect in studies of secondary production. An excellent review of the problem is given by Prepas and Rigler (1978). The Lanao sampling was specifically designed to avoid any serious weaknesses in the primary data base which might greatly reduce the scope for data interpretation.

The zooplankton samples were taken between September 1970 and November 1971 as part of a comprehensive study of the Lake Lanao plankton system. The zooplankton sampling program included the following: (1) A weekly vertical series of samples (5-m increments) taken at the index station (station 1, see Fig. 2–1). (2) A weekly series of three vertical tows, also taken at station 1, and an identical series of tows at station 2, 1 km distant (Fig. 2–1). (3) Monthly transects including duplicate tows at each of nine stations 1 km apart as indicated in Figure 2–1. (4) Special sampling programs to test particular hypotheses as needed.

The vertical series samples were taken with a self-closing, transparent Schindler–Patalas sampler (Schindler, 1969) equipped with a 35-μm mesh net. A very large trap was used (45.3 liters) to minimize escape of highly motile species and to provice sufficient numbers of rare species for counting. The trap was retrieved at a fast, constant speed with a motorized winch to prevent reopening, the major problem with traps of this type.

The vertical tows were taken with a metered townet calibrated each time it was used. The counts were individually corrected for filtration efficiency. Efficiency averaged very near 50% for the entire study period.

Since one of the townet sample series was always taken at the same time and place as the vertical series with the Schindler–Patalas (S–P) trap, a statistical comparison of the two sampling methods is possible. The vertical series samples are summed over all depths and the mean number per liter of each zooplankton type is obtained. This is compared with the mean of the replicate vertical tows at the same station. If both methods are good, then correlation between the vertical series and vertical tow means should be strong and the relationship between them should be essentially the same for all zooplankton types.

Scatterplots of abundance measured by townet versus abundance measured by S–P trap showed that a clumping of abundances in the lower ranges would require log transformations in many cases prior to correlation analysis. Log transformation was satisfactory to normalize the data whenever transformation was needed.

Table 5–1 lists some major zooplankton types and the results of the correlation analysis on these types. Except for calanoid adults and *Chaoborus*, all abundances required log transformation. All correlations are very strong and suggest that the slight departure from perfect correlation can be mainly accounted for by counting variance and expected replicate variance at a single

Table 5–1. Correlation Coefficients (r) and Slopes (a) for Best Linear Fit to the Relationship between Abundance in a Vertical Tow (X) and Abundance in a Vertical Series (Y) When the Tow and the Series Are Taken at the Same Time and Place[a]

Species/stage	r	a
Cyclopoids		
Nauplii	0.93	1.02
Copepodids	0.83	0.94
Adults	0.82	0.97
Calanoids		
Nauplii	0.91	0.85
Copepodids	0.93	0.86
Adults	0.80	1.85[b]
Conochiloides	0.94	0.94
Diaphanosoma	0.85	0.91
Bosmina	0.94	0.88
Chaoborus	0.71	1.60[b]

[a] Non-normal variables were log transformed prior to analysis. All variables except two (see footnote b) required transformation. All relationships are highly significant ($P \ll 0.01$, $n = 53$).

[b] Value for untransformed variables, not comparable to other a values.

sampling site on the lake. The only possible exception to this is *Chaoborus*, which is much more motile than the other species.

The slopes (*a*) for best linear fit to the transformed variables are very nearly uniform ($\simeq$ 0.95), suggesting that all species are captured with equal effectiveness by the two methods. The standard error of the transformed slopes is always about 0.04. Table 5–1 provides some basis for speculation that a detectable but small percentage of calanoids successfully avoids the S–P trap. The effect is not large enough to be of any interest, however. The value of *a* for untransformed variables should be very near 2.0, which corresponds to the mean filtration efficiency of 50% as measured by the meter on the townet, or to an a value of about 1.0 on transformed variables. The determination of best fit does produce the expected *a* values, except for *Chaoborus*, and this confirms the impression that escape in addition to that anticipated by the measure of townet filtration efficiency did not occur. *Chaoborus* produces an *a* value slightly lower than expected, and the standard error of the slope (0.10) indicates that this is not a chance variation. The implication of the lower *a* value for *Chaoborus* is that the S–P trap is 80% efficient on *Chaoborus*. This is probably explained by the escape of some *Chaoborus* through the bottom of the S–P trap before the lower door closes.

In the computations of abundance the S–P trap is assumed to be 100% efficient for all species and stages except *Chaoborus*, for which it is assumed to have 80% efficiency. The townet is assumed to have 100% efficiency for the volume of water actually filtered as indicated by the meter on the net. These assumptions are well strengthened by the foregoing analysis.

Counting

All samples were preserved immediately with Lugol's solution. At the time of counting, the sample was brought to constant volume (90 or 45 cm³, depending on densities) and subsampled (1 cm³) with a wide-bore automatic pipet which was filled and emptied vigorously several times prior to sampling in order to disperse the specimens in the sample. Comparisons of the numbers of zooplankton in a subsample with the total number in a sample showed that the subsampling technique produced an unbiased estimate of the mean and could therefore be considered a representative random sample (Cassie, 1971). This was true for all species except *Chaoborus*, small shrimp, and fish fry, which did not subsample adequately. Consequently abundances of these items were always estimated on the basis of total counts rather than subsamples.

The subsamples were placed in a tray with slots of exactly 1 cm³ capacity. Counting was done with a stereomicroscope at 100× for small species and 64× for larger ones. No fewer than two subsamples were ever counted for any species or developmental stage. For a species or developmental stage whose count had not exceeded 100 by the end of the tabulation on the second subsample, additional subsamples were counted up to a total of four. On occasions when the calanoid adults were rare, the entire sample was counted for them, but this was not practical for the smaller species.

Table 4–1 gives a complete list of the species and stages that were distinguished in the counts. The table also shows the mean abundance of each species and stage over the study period and the number of individuals counted in an average week to obtain the abundance estimates for that week. Separation of stages differed some between the taxa according to life history patterns. Major differences were as follows.

1. Copepods

As indicated in Table 4–1, it was possible to separate all of the developmental stages of both the copepod species. This is usually impossible due to the similarity of the developmental stages among cyclopoids or calanoids of different species. Since only two very different species needed to be distinguished in this case, there was no possibility of interspecific confusion. Even so, distinctions between naupliar stages of a given species were difficult to make and required construction of keys based on animals reared in the laboratory. Although it was not possible to rear either species all the way through development, the keys for both species were verified through nauplius 5, which includes the most difficult stages.

In temperate lakes, advanced copepodid stages of cyclopoids often undergo diapause in the mud (Hutchinson, 1967). This can create sampling bias. In Lake Lanao, copepodids were never found in the mud. This is consistent with findings for other warm lakes in which cyclopoids remain planktonic throughout the life cycle (Ravera, 1954; Fryer, 1957a; Burgis, 1971). The counts are thus assumed to represent all developmental stages fairly.

Adults of both species were distinguished on the basis of number of urosomal segments and secondary sexual characteristics. By these criteria, adult males always outnumber females, but this does not necessarily imply that the sex ratio of actively breeding individuals is similarly out of balance. There may be a difference between sexes in speed with which phenotypic evidence of sex first appears.

2. Cladocera

No distinction was made between *Diaphanosoma modigliani* and *D. sarsi* in the counts. For all cladocerans, eggs and embryos were tabulated separately but are usually merged in the data analysis because the duration of development is very short.

3. Rotifers

All adults were distinguished in the counts, as were the eggs of the two *Keratella* species and *Tetramastix*. The eggs of *Conochiloides, Hexarthra, Polyarthra,* and *Trichocerca* were not distinguished because of their similarity and their tendency to become detached from the adult. The number of eggs attributed to each of these four species is always assumed to be proportionate to the relative abundance of adults. Because of the strong numerical dominance of *Conochiloides* among these four species, the pooled egg approximation does not have significant adverse effects on the data analysis.

4. Chaoborus

Individual *Chaoborus* were always measured to the nearest 100 μm. A size-frequency distribution therefore supplements the total abundance data on all dates.

Size and Biomass Measurements

Volumes were computed for all of the species and developmental stages listed in Table 4–1. Mean length measurements, which exclude antennae and setae, are recorded in Table 4–1. From the mean lengths and other mean dimensions, clay scale models were constructed, and the volumes of these models were determined by water displacement. The results are listed in Table 4–1.

For both calanoids and cyclopoids, certain groups of stages proved to have common length–weight relationships which are useful in the conversion of individual lengths to weights. The equations describing these relationships are shown in Table 5–2. The equations predict the volumes of scale models within about 5% of the measured volumes. It is clear from the equations that the cyclopoids change shape between N6 and CI, as might be expected. The nauplii are rounder and thus increase in weight faster for a unit of length. The N1 stage does not fit the general pattern, and a separate slope for N1 cannot be calculated due to its narrow range of lengths, so this stage is omitted from the equations. The N1 stage is slimmer than other stages as it emerges from the egg, and this accounts for its divergence from the line. The male cyclopoid also has a slimmer shape than the female and the copepodids, and thus is omitted from the general equations for cyclopoids.

Tropodiaptomus nauplii have a shape very similar to *Thermocyclops* nauplii. The N1 stage is not divergent, however. As with *Thermocyclops,* a change in shape occurs between N6 and CI, but the lengthening process ex-

Table 5–2.   Relation of Wet Weight to Body Length in the Developmental Stages of Cyclopoid and Calanoid Copepods from Lake Lanao

Taxon/stage	Relation of length (μm) to wet weight (μg $\times$ 10^6)
Thermocyclops	
N2–N6	$\log (Y) = 2.416 \log x + 0.361$
C1–C5, ♀	$\log (Y) = 2.188 \log x + 0.803$
Tropodiaptomus	
N1–N6	$\log (Y) = 2.304 \log x + 0.612$
C1–C3	$\log (Y) = 1.086 \log x + 3.84$
C4–C5, ♂, ♀	$\log (Y) = 2.374 \log x + 0.295$

tends over three stages (CI–CIII) rather than one. The last copepodid stages and adults are uniformly more rounded and the male is not markedly divergent in shape.

The length–weight relation for *Chaoborus* was already known (Lewis, 1975). Table 4–1 reports mean total length for the year. Because *Chaoborus* varies so greatly in size (4–300 μg wet weight), the mean weight in Table 4–1 is computed as the average of all individual weights rather than as the weight of an individual of average length.

Chapter 6

Spatial Distribution of the Zooplankton

The analysis in the chapters which follow is conducted as if the zooplankton community in the vicinity of stations 1 and 2 (Fig. 2–1) were typical of the entire plankton environment of Lake Lanao. As indicated in Chapter 5, the extended time series of weekly samples is an average from two stations 1 km apart in order to minimize the effect of local horizontal spatial variation. In addition, multiple samples were taken at each station each week to minimize any effect of patchiness on very small distance scales. The final averages on which all the dynamic analysis is based thus typify specifically that portion of the lake near the 50-m contour where stations 1 and 2 are located (Fig. 2–1). The extent to which this portion of the lake is representative of the entire plankton environment can be judged on the basis of the following summary of spatial analyses which have been developed more extensively in previous publications (Lewis, 1975, 1977a, 1978c, 1979). In addition, the vertical distribution of the zooplankton is worthy of some brief treatment here.

Vertical Distribution

In the vicinity of stations 1 and 2 at least some depletion of oxygen concentrations below a depth of about 20 m is quite common during the stratification season, especially when the development of temporary thermoclines restricts active mixing to the uppermost part of the water column (Fig. 6–1). Complete oxygen depletion is of course most likely near the mud–water in-

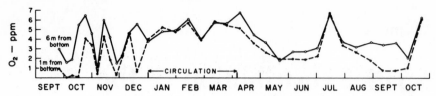

Figure 6–1. Variation in the oxygen concentration near the bottom of Lake Lanao at station 1, near the 50-m contour, showing the changes in suitability of the lower water column for oxygen-requiring zooplankton (1970–71).

Table 6–1. Weighted Mean Position in the Water Column of Lake Lanao Zooplankton from Station 1 Near the 50-m Contour[a]

Species/stage	Weighted mean depth (m)		Day–night (m)
	Day	Night	
Copepods			
Thermocyclops nauplii	21.6	18.6	3.0
Thermocyclops copepodid/adult	19.2	16.6	2.6
Tropodiaptomus nauplii	18.5	16.0	2.5
Tropodiaptomus copepodid/adult	21.6	17.6	4.0
Cladocera			
Diaphanosoma	19.8	13.4	6.4
Moina	19.0	18.7	0.3
Bosmina	20.5	14.2	6.3
Rotifers			
Conochiloides	16.5	14.2	2.3
Hexarthra	19.3	12.2	7.1
Polyarthra	16.8	16.4	0.4
Keratella procurva	16.8	11.2	5.6
Keratella cochlearis	20.3	19.0	1.3
Trichocerca	17.2	13.3	3.9
Tetramastix	16.8	16.8	10.0
Chaoborus			
Instar I	21.7	17.6	4.1
II	31.8	21.1	10.7
III	40.3	16.8	23.5
IV	42.6	19.9	22.7

[a] Data are averages based on vertical profiles taken weekly between September 1970 and October 1971. Daytime samples were taken between 0930 and 1130 *h* and night samples were taken between 1900 and 2000 *h*.

terface. Because of the potential for oxygen depletion and drastically reduced phytoplankton abundances below the circulating layer (Lewis,
1978e), the deeper portion of the water column is unsuitable as a permanent
habitat for zooplankton herbivores in Lake Lanao. Most of the zooplankton
species do show some tendency to seek deeper water during the day,
however, as indicated in Table 6–1. Vertical migration is of course extremely widespread among the freshwater zooplankton (Hutchinson, 1967),
and a general tendency toward nocturnal upward movement such as that indicated in Table 6–1 is by far the most common pattern.

The difference in migration amplitudes between zooplankton species and
developmental stages imparts a definite vertical spatial organization to the
zooplankton community which probably is of some ecological significance.
For example, the *Chaoborus* population is vertically arranged during the
daytime in a remarkably uniform pattern resulting from a close relationship
between mean depth in the water column and body size (Fig. 6–2). This daytime pattern probably results from a balance between the advantage of seeking highest prey densities toward the middle and upper portions of the water

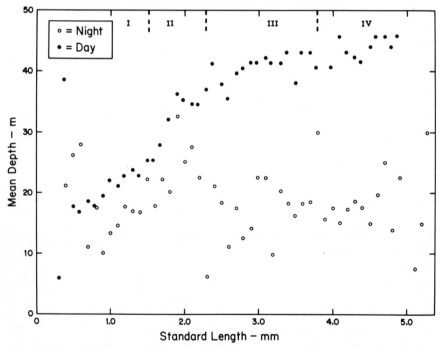

Figure 6–2.   Weighted mean annual depth on the water column of *Chaoborus* larvae
as a function of size. Daytime data are based on 47 weekly or biweekly abundance–
depth profiles and nighttime data are based on 20 biweekly profiles. Instar divisions
are shown on the upper abscissa. Standard length times 2.0 equals total length.

column and the advantage of seeking the best refuge from fish predation in the lower portion of the water column (Lewis, 1975, 1977a). Since the largest organisms are subject to more intense predation from fish, the balance of these two pressures would dictate a distribution of size with depth similar to that shown in Figure 6–2. At night the *Chaoborus* spatial structure breaks down, as indicated in Figure 6–2, presumably because of the reduced danger from predators that rely on visual cues.

The herbivore species all have migration amplitudes which are less than that of large *Chaoborus,* but most do show a considerable difference between day and night distributions (Table 6–1). There is some variation through time in the migration amplitudes of individual species and stages. The daytime position of *Chaoborus,* for example, is statistically related to the oxygen profile (Lewis, 1975). The basic migratory pattern indicated in Table 6–1 persists, however, despite some variations in amplitude and dispersion.

Multiple factors probably account for daily redistribution of the community. Predation is strongly indicated in a number of studies as a general evolutionary cause for migration (Zaret and Suffern, 1976). Predation is relevant to vertical distribution of Lake Lanao herbivores because of the heavy losses of herbivores to *Chaoborus.* Similarly, the losses of *Chaoborus* to fish predation must be related to vertical distribution of *Chaoborus.* Vertical separation enforced by resource-based competition of the type discussed by Lane (1975) may also contribute to the differing vertical distributions and migration amplitudes, but this is difficult to demonstrate. Vertical spatial segregation of species is not very great among the herbivores (Table 6–1), and herbivores show little evidence of taxing the food supply, as will be demonstrated in the analysis of community dynamics.

Horizontal Variation: Fixed Patterns

Horizontal spatial variation can be divided into a fixed component and an ephemeral component. These components have been separated for the Lake Lanao zooplankton by analysis of variance procedures (Lewis, 1978c). The fixed component of horizontal spatial variation is associated with temporally stable differences between stations. Such temporally stable differences might result from fixed abundance gradients of various kinds or from fixed patches. In contrast, the ephemeral component of horizontal spatial variation (so named by Platt and Filion, 1973) is a space–time interaction, i.e., ephemeral horizontal variation is horizontal patchiness which changes randomly through time.

Because Lake Lanao does not have any major point sources of pollution or other important kinds of externally enforced heterogeneity, the fixed component of spatial variation is considerably less than the ephemeral component. A statistically detectable and quantifiable fixed component of spatial variance does exist, however. A quantitative examination of this fixed component of horizontal spatial variation for Lake Lanao showed that it is caused by temporally stable trends in the species composition of the zooplankton community related to depth of the water column (Lewis, 1978c). Although these trends are moderate in magnitude, they are biologically significant and are also revealing of the processes which control community structure in Lake Lanao.

The trends which showed up in the studies of zooplankton horizontal spatial distribution are expressed in relative terms in Figure 6–3. Cyclopoids of all developmental stages show a strong positive relationship between abundance per unit area and depth of the water column, as do all four instars of the primary carnivore *Chaoborus*. All of the other zooplankton, including

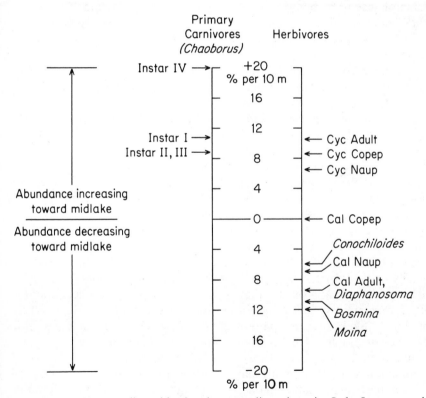

Figure 6–3. Temporally stable abundance gradients in major Lake Lanao zooplankton species and stages. Species and stages are arranged on a scale of percentage increase or decrease in abundance per unit area for each 10-m change in the depth of the water column between the 25-m contour and midlake (110 m). Herbivore gradients can be explained on the basis of primary carnivore gradients.

calanoids, cladocerans, and rotifers, show negative depth–abundance relationships (i.e., they decrease in areal abundance toward midlake). The abundance gradients range in intensity but are especially strong for cladocerans (Fig. 6–3).

Since no temporally-stable nutritional gradients exist for herbivores in the limnetic zone, mortality mechanisms are implicated as causes for the gradients. The main herbivore mortality is attributable to *Chaoborus*. The feeding rates, abundances, and feeding selectivities of all *Chaoborus* instars were used to calculate potential predation losses as a percentage of the stock of each prey type assuming an average community structure for the herbivore prey. These computed losses had a significant slope when regressed against depth, indicating the existence of predation gradients for individual herbivore species and developmental stages (Lewis, 1979). The predation gradients showed statistically significant negative relationships to the observed abundance gradients of herbivores. Strong evidence was thus established for the maintenance of herbivore abundance gradients by predation. The predation of primary carnivores *(Chaoborus)* on herbivores appears to be responsible for an unexpected amount of fixed pattern in zooplankton community structure.

If patterns in community structure of herbivores can be explained on the basis of patterns in primary carnivore distribution, some explanation must be sought for patterns in the distribution of primary carnivores. As will be shown in subsequent chapters, *Chaoborus* mortality is extremely heavy and can be accounted for almost entirely by fish predation. As one moves from the edges of the plankton environment in some 20 m of water toward the middle of the lake where the water column is over 100 m deep, the extent and quality of the deep-water refuge for *Chaoborus* becomes greater. In shallower water, an anoxic or low-oxygen refuge that might be difficult for fish to exploit is present only over short portions of the year. More central portions of the lake offer such a refuge from fish predation for a greater portion of the year and also offer water of greater depth, which may itself be valuable protection from fish. The gradient of conditions appears to be especially critical for the largest *Chaoborus*. Instar IV *Chaoborus* are extremely rare as one approaches the littoral zone and there is a generally increasing abundance of all *Chaoborus* instars toward midlake (Fig. 6–3). It seems highly likely that this pattern is established and maintained by fish predation, which falls most heavily on advanced *Chaoborus* stages.

In summary, fixed horixontal spatial variations do occur in the structure of the zooplankton community in Lake Lanao. The year-round average structure at any particular location in the lake will thus depend on the depth of the water column at that location. A dynamic analysis based on samples taken around the 50-m contour is typical only in the strictest sense of events which occur approximately midway along the gradient of community structure with depth. The factors affecting community structure and energy flow midway along the community structure gradient are likely to be in principle the same throughout the community, however.

Horizontal Variation: Ephemeral Patterns

The ephemeral component of spatial variation has been generally emphasized in plankton biology because the establishment of moving water masses with discrete chemical and biological characteristics seems intuitively to be the most likely source of variation in the horizontal plane. In Lake Lanao, ephemeral horizontal variation is indeed more important quantitatively than is the fixed component of variance, but it remains to be seen whether the biological significance of the ephemeral component is really greater.

The most intuitively meaningful expression of ephemeral variation is a relative one. Table 6–2 shows ephemeral horizontal spatial variation in the abundance of zooplankton herbivores in relation to temporal variation as determined by a components of variance procedure (Lewis, 1978c). The components of variance procedure merely uses analysis of variance methods to split up the total variance in a space–time data matrix into (1) variance through time, (2) spatial variance that is stable (fixed) through time, (3) spatial variance that is changing (ephemeral) through time, and (4) error variance. Table 6–2 simply reports the ratio of ephemeral spatial variance (com-

Table 6–2. Ratio of the Ephemeral Component of Horizontal Spatial Variation to Total Temporal Variation[a]

Species/stage	Ratio of ephemeral spatial variance to total temporal variance
Copepods	
Thermocyclops nauplii	0.18
Thermocyclops copepodids	1.02
Thermocyclops adults	1.33
Tropodiaptomus nauplii	0.74
Tropodiaptomus copepodids	0.56
Tropodiaptomus adults	0.67
Cladocerans	
Diaphanosoma	2.43
Moina	3.77
Bosmina	0.25
Rotifers	
Conochiloides	0.48

[a] Variance components were computed from numbers of organisms per unit area in samples taken over the entire range of annual conditions at multiple stations transecting the lake.

ponent 3) to temporal variance (component 1) to show the relative importances of the two.

Ephemeral spatial variation is considerably less than total temporal variation for most Lake Lanao herbivores. The large cladoceran species *(Diaphanosoma, Moina)* are a striking exception, however, as the extent of ephemeral variation is much greater in these species. The differences between species in relative importance of ephemeral spatial variation may be related to differences in adaptive strategies for existence in the plankton zone (Lewis, 1978c).

Replicate sampling at each of two stations 1 km apart provides the basis for dynamic analysis, as indicated in Chapter 5. Use of weekly averages from such a sample set reduces the interference of ephemeral spatial variation in the analysis of dynamics. In addition frequent sampling over an extended time interval reduces the relative importance of ephemeral spatial variation in obscuring temporal trends. Ephemeral spatial variation does add to the background noise in the data set, however, and may thus obscure very weak temporal trends.

Overview of Horizontal Variation

So little quantitative information on horizontal spatial variation is available that it is difficult to generalize about the relative importance of horizontal spatial variation in the Lake Lanao zooplankton community as compared with other temperate and tropical plankton communities. Clearly the lake is large enough to develop some significant patchiness but not so large that discrete water masses with lengthy independent histories are routinely established, as they often are in marine environments (Parsons et al., 1977). The action of wind on the upper water column maintains a plankton environment of surprisingly great homogeneity in Lake Lanao.

The fixed horizontal patterns in community structure within the plankton zone were unexpected but provide a welcome means for independent confirmation of the important role of predation in controlling community structure, as will become evident in later chapters. Several cases of fixed horizontal pattern in a single species based on predation have been documented (Green, 1967; Zaret, 1972; Kerfoot, 1975), but temporally stable gradients in the structure of entire communities based on water column depth and maintained by predation have apparently never been quantitatively demonstrated in other lakes, although numerous investigators may have suspected their existence.

Ephemeral variation, although studied to a much greater extent in various community types (Margalef, 1958; 1967; Platt et al., 1970; Platt and Denman, 1975; Powell et al., 1975; Wiebe, 1970), varies enormously according to the

physical and chemical characteristics of the habitat and thus resists generalization. Certainly many freshwater and marine systems are patchier in an
ephemeral sense than Lake Lanao simply because the conditions in Lake
Lanao act against the development of enduring isolated subsystems in the
plankton zone. The exact biological importance of ephemeral spatial variation in the Lake Lanao system or in zooplankton communities in general is
extremely difficult to judge even on the basis of quantitative information. It
seems likely, however, that both herbivores and predators are adapted to the
existence of patchiness in plankton environments and that evidence will
eventually be found of such adaptations in the demographic characteristics,
morphology, or behavior of zooplankton organisms.

Chapter 7

Zooplankton Abundance Trends

Figures 7–1 and 7–2 give an overview of temporal abundance patterns in Lake Lanao zooplankton. Figure 7–1 shows how the three major trophic levels respond to major seasonal and nonseasonal variations in abiotic factors.

Total phytoplankton biomass as shown in Figure 7–1 is very closely correlated with phytoplankton production, which is in turn under the control of nutrient availability, light availability, and turbulence. The seasonal circulation period is unfavorable for phytoplankton because of the depth of circulation, which limits light availability. The seasonal biomass depression is evident in Figure 7–1. At other times of the year (e.g., June 1971) declines in phytoplankton biomass are associated with depletion of inorganic nitrogen (Lewis, 1974). It is clear from Figure 7–1, however, that the degree of annual fluctuation in total phytoplankton biomass is much less than would be expected in temperate lakes and that substantial amounts of phytoplankton biomass are present even under the worst conditions of light limitation or nutrient depletion.

Figure 7–1 also shows that total herbivore biomass seldom exceeds total phytoplankton biomass. The annual mean ratio of phytoplankton to zooplankton biomass (P/Z ratio) is 1.47. This is in the upper portion of the range reported by Ruttner (1937) for a series of Alpine lakes (0.07–2.1).

An extended series of phytoplankton–zooplankton comparisons was done by Nauwerck (1963) on Lake Erken, Sweden. The mean P/Z ratio he reports (0.18) is far below the ratio for Lake Lanao, suggesting some major contrasts with Lanao. Actually the differences are probably not nearly so great as they appear, as Nauwerck's values for zooplankton biomass seem to be overesti-

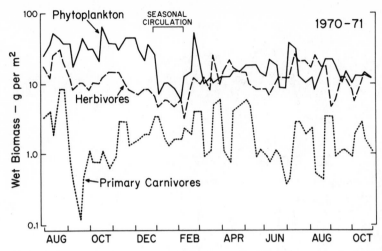

Figure 7–1. Abundance of phytoplankton, total herbivores, and primary carnivores at weekly intervals in Lake Lanao. Primary carnivore curve has been slightly smoothed (moving average of 14 days).

mates. *Eudiaptomus graciloides,* which accounts for more than half the standing crop in Erken, is estimated at 2×10^8 μm^3 per adult, or 200 μg wet weight, by Nauwerck. Although the lengths are not given, the same species in Lake Esrom ranges between 575 and 625 μm (head and thorax only, Bosselmann, 1975). Assuming an average of 600 μm for adults, the curves of Bottrell et al. (1976), which agree with the data of Table 4–1, would indicate a wet weight of only about 35 μg for *E. graciloides,* or about one-sixth of the value Nauwerck used in his estimates. Nauwerck's copepodid and cladoceran weights seem similarly high, so the true P/Z ratio is probably much higher than his estimate. Unfortunately, too few exhaustive studies of the type done by Nauwerck on Erken are available to show whether the P/Z ratio of Lanao is unusual or not.

Figure 7–1 shows that the ratio of phytoplankton to herbivore biomass in Lake Lanao is only infrequently less than 1.0. In addition, phytoplankton biomass turnover always exceeds herbivore turnover by a wide margin. Together these facts suggest that Lanao herbivores may leave major portions of the phytoplankton uneaten. Approximations of grazing rates of Lanao herbivores (Lewis, 1978b) have in fact shown that the herbivores probably do not eat more than 10% of the total annual primary production. There is nevertheless some correspondence between the amount of herbivore biomass and the amount of phytoplankton biomass, suggesting that phytoplankton density or quality affects herbivore growth even though the herbivores do not ingest the majority of primary production. This matter will be considered in some detail in chapters dealing with the growth control of herbivores.

Although changes in herbivore abundance tend to be less abrupt than

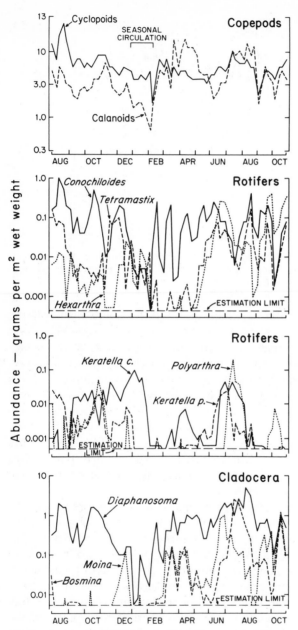

Figure 7–2. Abundance of all herbivore species except the rarest rotifer species (*Trichocerca*) at weekly intervals over the study period. Lines run just above estimation limit marker when individuals were counted in such low numbers as to make abundance estimation unreliable. Complete absence of a species from samples is indicated by termination of its line at the estimation limit.

changes in phytoplankton abundance owing to the greater amount of inertia in the herbivores, the suboptimal conditions of the seasonal circulation period are definitely reflected in the herbivores as well as in the phytoplankton (Fig. 7–1). This is not so true, however, of the periods of nutrient depletion, which place the phytoplankton community under great stress but do not always result in a reduction of herbivore biomass.

The primary carnivore level in the open water plankton community of Lake Lanao is limited almost entirely to *Chaoborus*. Figure 7–1 shows the variation in abundance of *Chaoborus* in comparison with the abundance of its herbivore food items. There is no identifiable relationship whatever between the abundance of *Chaoborus* and herbivores, suggesting that carnivores are not particularly sensitive to variation in food abundance. This also will be considered in some detail later.

Figure 7–2 gives more detailed information on the temporal trends in herbivore species. All the Lake Lanao species except the rarest rotifer *(Trichocerca)* are represented in the figure.

It is clear from Figure 7–2 that the two copepod species have closely related ecological requirements. The cyclopoids in particular maintain a remarkably steady biomass over the course of the year. The exact reasons for peaks and declines in either species are not obvious and will require statistical evaluation, however.

All of the major rotifer species are in general adversely affected by the onset of seasonal circulation, but *Conochiloides,* which is easily the dominant rotifer, shows full recovery immediately following full circulation whereas the other species do not. There is a pronounced instability in the *Conochiloides* population immediately following recovery from circulation, however, which is not characteristic of the population the rest of the year. All of the major rotifer species thrive during periods of severe nutrient depletion when phytoplankton biomass is on the decline.

Among the rare rotifers, *Keratella cochlearis* is anomalous in failing to show an immediate negative response to seasonal circulation. The population patterns of the *Keratella procurva* are considerably different despite its morphological similarity to *K. cochlearis.*

Among the cladocerans, *Diaphanosoma* is unique in its ability to maintain relatively high abundances at all times of the year except during the circulation period. The other two species pass through extended periods of extreme rarity but at certain times of the year show pronounced abundance peaks.

As described in Chapter 3, the phytoplankton community passes through a successional sequence several times per year in connection with changes in depth of mixing. The complexity of the biomass variations in herbivore species is such that it is impossible to tell by simple inspection of Figure 7–2 whether there is any corresponding successional trend in zooplankton herbivores. Statistical analysis of the factors which control growth and reproduction will allow consideration of successional sequences.

Chapter 8

Zooplankton Development Rates

Cyclopoids

Burgis (1970, 1971) has studied the development time for eggs of *Thermocyclops hyalinus* from Lake George, Uganda, in considerable detail. Since pilot experiments on the development of eggs from Lanao indicated very little difference between the Lake Lanao and Lake George populations in this respect, no comprehensive experiments were necessary. Six independent laboratory determinations based on the observation of groups of three ovigerous females in graduate cylinders indicated a mean development time of 47.7 h at 23°C for the Lanao population. Given the mean development rate reported by Burgis (1970) and the linearity of the time reciprocal with temperature over the range 20–30°C, the Lake George population would be expected to show a development time of 44.6 h under similar conditions, and the generalized relationship given by Schindler (1972) would predict a development time of 51.5 h. These estimates are in sufficiently close agreement that the differences could easily be due to experimental techniques. The temperature–development rate relation of Burgis is therefore adopted here.

The variation of temperature with depth and time has been a serious source of error in studies of zooplankton production (Prepas and Rigler, 1978). Fortunately, Lake Lanao varies little in temperature. The mean temperature of the water column at station 1, excluding the top 2 m, in which there is a large diel heat exchange, ranged over the study period between 24.3°C (March 1971) and 26.5°C (September 1970). Although the animals migrate through a thermal gradient at some times of the year, the gradient is typically less than 2°C and is most often less than 1°C (Lewis, 1973a). This vertical gradient will be ignored in the calculations as it is unlikely to cause an error of more than 5% in the development rate calculations. The seasonal maximum and minimum temperatures averaged over the water column correspond to egg development times of 40.0 and 34.3 h. Thus a development time of 37 h based on annual mean temperature would be in error by less than 10% at the extreme, and can be used in all computations.

The mean egg size reported in Table 4–1 (57 μm) suggests that the eggs of the Lanao population were smaller than those of the Lake George population (76 μm, Burgis, 1971). This difference is only apparent, as the Lanao measurement is based on the diameter of the yolky portion of the egg. The outer membrane separates from the yolk (also noted by Burgis, 1970), and thus increases the apparent diameter of the egg. If the membrane is measured rather than the yolk, then the mean for the Lanao population is 72 μm, which is much closer to the diameter reported by Burgis for Lake George.

Laboratory rearing of *T. hyalinus* was possible for short periods but not for long periods. Even if rearing were possible, it would provide an unreliable estimate of development times for the instars because realistic conditions are almost impossible to maintain. A cohort analysis was therefore adopted as the means for estimating the length of developmental stages beyond the egg. The analysis was done by a method similar to that of Rigler and Cooley (1974), as follows.

1. Distinctive cohorts were identified (Fig. 8–1). Only cohorts spanning several weeks were used so that the sampling interval would have minimum effect on the analysis.

2. The temporal "center of mass" for a cohort (i.e., its time average weighted by abundance) was determined as follows:

$$T_k = \sum_{i=f}^{l} a_{i,k} t_{i,k} \Big/ \sum_{i=f}^{l} a_{i,k}$$

where f is the number of the first time interval spanned by the cohort at stage k (1 interval = 7 days), l is the number of the last interval spanned by the cohort at stage k, $a_{i,k}$ is the abundance of stage k in interval i, $t_{i,k}$ is the number of the ith interval spanned by stage k, and T_k is the location along the time axis of the center of mass for the cohort.

3. The time between midpoints of developmental stages was computed as the difference between T_k for various values of k.

4. Rigler and Cooley made separate calculations of the duration of each instar. For the Lanao data, instars were grouped as follows for purposes of calculation: N1–3, N4–6, CI–III, CIV–V. This is more practical due to the fast pace of development and sharpness of peaks of Lanao populations. The differences in T_k values for two of these com-

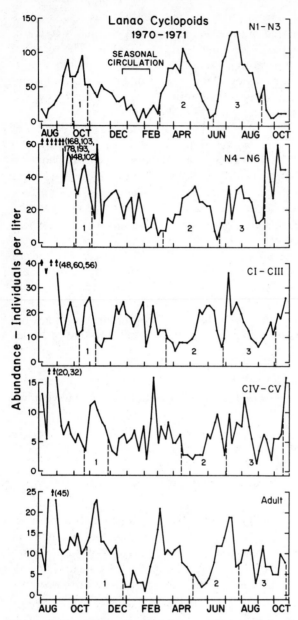

Figure 8–1. Changes in abundance with time of developmental stages of *Thermocyclops hyalinus* in Lake Lanao. The three cohorts selected for life history analysis are indicated with vertical dashed lines.

posite groups is taken as the time required for development from the temporal midpoint of one composite to the temporal midpoint of the next. Thus if $k = 1$ represents composite N1–N3, and $k = 2$ represents N4–N6, then $T_2 - T_1$ is an estimate of development time over the interval N2.5 → N5.5, whose midpoint is the beginning of instar N4. The interval is 3 instars long, thus the duration of instar 4 is computed as $(T_2 - T_1)/3$. This assumes that adjacent instars require similar development times. The more complex iterative procedure of Rigler and Cooley avoids this assumption to some extent, but introduces other problems arising from uneven mortality. Since it can be shown in the laboratory that adjacent developmental stages do in fact generally have very similar development times, the simplification seems justified.

5. Development time is plotted against instar number (Fig. 8–2) and a smooth curve is drawn through the points. Development times of all instars can be read from this curve. Because instar N1, which does not

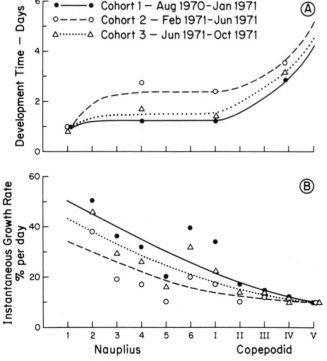

Figure 8–2. (A) Development time plotted against instar for cyclopoid copepods. Separate curves are drawn for three cohorts representing different portions of the study period. (B) Growth rate per day plotted against instar.

feed, often has a development time much shorter than other naupliar stages (Comita, 1972; Rigler and Cooley, 1974), newly hatched N1 cyclopoids were observed in the lab so that a separate estimate could be made for this instar. Development time averaged 0.9 days, and this is added to the graph in Figure 8–2.

As indicated by Rigler and Cooley, the computations are subject to certain errors, particularly if mortality is uneven through time for a particular cohort. The method nevertheless provides the most rational reconstruction of the life history of copepod field populations.

The general life history pattern of *Thermocyclops* in Lanao is similar to that of the same species in Lake George insofar as the naupliar stages are shorter than the copepodid stages (Burgis, 1971). The developmental pattern for the Lanao population is also very similar to the one documented by Rey and Capblancq (1975) for *Mixodiaptomus* in a coldwater lake, as the stages are of nearly equal duration until CIII.

Burgis (1971) reports development times for *Thermocyclops* in Lake George of approximately half those given in Figure 8–2 for the Lanao population. Individuals from the Lake Lanao population are 5 to 10% shorter and weigh at least 20% less than *Thermocyclops* from the Lake George population (Table 4–1 and Burgis, 1974), so the differences in development rate cannot be explained on the basis of size differences. Thus while the life history pattern for *Thermocyclops* appears to be rather standard, development rates seem to differ considerably even between lakes of similar temperature.

Burgis determined development rates by two techniques. Animals were reared in the laboratory to provide one estimate, and a second estimate was derived from the ratios of numbers of individuals in various stages using the assumption that mortality was uniform across stages. Both methods could be questioned, but their agreement tends to confirm her estimates. Cohort analysis was not possible for Burgis because the population was too stable. In Lake Lanao, the assumption of equal predation across instars is impossible owing to marked selection of copepodids over nauplii by *Chaoborus,* the dominant predator (Lewis, 1977a). In fact the ratio of nauplii to copepodids is approximately 4 : 1 in Lanao, whereas Burgis reports a ratio of approximately 1 : 1 in Lake George. The predation pressures on the *Thermocyclops* populations in these two lakes thus appear to be very different and require different approaches in the determination of development rates.

Slower development of Lanao *Thermocyclops* probably cannot be attributed to temperature even though Lake George is slightly warmer than Lanao. Food is one possible explanation, and a second is the operation of selective pressures to mold the life cycles to the individual characteristics of the lakes, thus causing divergence. These factors cannot be separated here.

Development times of the three *Thermocyclops* cohorts shown in Figure

8-2 differ considerably. Although zooplankton development rates are for purposes of calculation commonly assumed to be stable through time, variations in development rate are quite likely to occur in nature as conditions change in a lake. Some portion of the variation between cohorts in Figure 8-2 may be due to inaccuracies in the assumptions and calculations, but the curves do not cross each other and are thus internally consistent and suggestive of actual changes in development rates between cohorts rather than a random pattern of inaccuracies caused by calculation errors.

The differences between cohort development rates will be assumed real and will thus be carried into the production computations. The three cohorts do not cover the entire study period, so some extrapolation is required. The study period is divided into three major growth periods (Fig. 8-2), and each cohort is assumed to typify the period within which it lies.

Given the duration of each instar (Fig. 8-2) and the mean weight of organisms in the instar (Table 4-1), it is possible to compute the growth rate of each instar for the three cohorts. The assumption is made that within each instar, individual growth is exponential (a satisfactory approximation for a short time interval). Thus for a time interval t (days),

$$g = \frac{\ln W_t - \ln W_0}{t}$$

where W_t is the weight of an average organism at the end of the interval t, W_0 is the weight at the beginning, and g is the instantaneous growth rate (proportion per day).

Values of g were determined from time intervals and weight changes corresponding to the spans between midpoints of adjacent developmental stages. Adjacent values of g were then averaged to obtain an approximation of the growth rate of the instar spanned by the two adjacent time intervals. The values of g, expressed as percentage/day, are shown in Figure 8-2. The calculations were separate for each cohort because the development times differ between cohorts. A smooth curve was drawn through the points. Growth rates used in later computations will be taken from this curve. The scatter of points about the curve is due principally to variance in weight estimations, especially within the smallest instars.

The computations involve several sources of error. Aside from the errors inherent in the determination of instar durations, the use of average weights ignores possible weight differences in a given instar on different dates, and any asymmetrical distributions of weight or weight gain within instars. Marked seasonal changes occur in the weights of given instars in temperate lakes, particularly if one cohort overwinters (e.g., Bosselmann, 1975). No marked body size changes occurred in the Lanao populations, however, hence assumption of constant weight for given instars is not as misleading as it would be for some temperate lakes.

Figure 8–2 shows that growth rate declines substantially in older instars, as might be expected, and levels off at about 10% per day.

Calanoids

The development time of *Tropodiaptomus* eggs was studied in the laboratory by a method identical to that used for *Thermocyclops*. The mean development time of the eggs at 23°C was 50.0 h. No extensive studies are available of the relationship between development time and temperature in tropical freshwater calanoid species. It will be assumed for present purposes that the relationship between development time and temperature is identical to that used for *Thermocyclops*. Given this assumption, the development time for calanoid eggs at mean lake temperature would be 38.6 h. This figure will be used in all the computations. The summary curve presented by Bottrell et al. (1976, Fig. 13) for composite calanoid data from the literature predicts a development time of 41 h and is thus in good agreement with the assumptions.

The duration of the developmental stages was determined by cohort analysis identical to that used for *Thermocyclops*. Figure 8–3 shows the three cohorts that were used in the analysis. Figure 8–4 shows the development times for all instars as obtained from the cohort analysis. The development time of the N1 stage was determined in the laboratory (0.7 days).

The points for cohorts 1 and 2 fit smooth curves whose shapes are much the same as the curves obtained for *Thermocyclops* (Fig. 8–2). Cohort 3, however, is irregular, and thus creates some problems in constructing the curve. The final approximation assumes that the curve shape is similar to that of other cohorts.

The development of *Tropodiaptomus* is unexpectedly rapid by comparison with the development of *Thermocyclops*. *Tropodiaptomus* is considerably larger, but nevertheless matures as rapidly or more rapidly than *Thermocyclops*. Although comparisons with other tropical calanoid populations are not possible, *Tropodiaptomus* in Lake Lanao develops at only a slightly faster pace than *Eudiaptomus* in Lake Erken (Nauwerck, 1963). Since *Tropodiaptomus* in Lake Lanao is smaller than *Eudiaptomus* in Lake Erken, and since *Tropodiaptomus* is growing at a considerably higher temperature, it is not unreasonable that the development time should be shorter. The rapid development of *Tropodiaptomus* in Lake Lanao is thus consistent with development of certain temperate populations. The ecological explanation for shorter development times in *Tropodiaptomus* than in *Thermocyclops* is not obvious.

Figure 8–4 shows that the calanoid growth rates decline with size, as do those of *Thermocyclops,* but the decline is not so steep and the curve shape differs, as there is a considerable drop in growth rate in the last instar. The

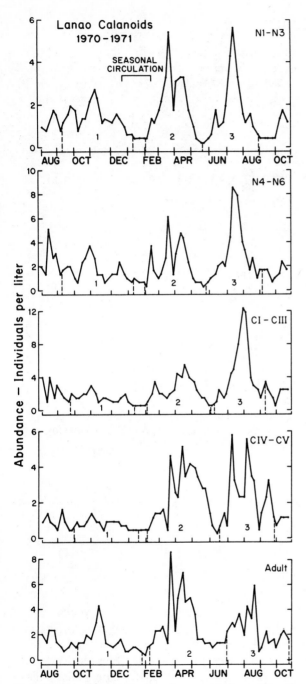

Figure 8–3. Changes in abundance with time of developmental stages of *Tropodiaptomus gigantoviger* in Lake Lanao. The three cohorts selected for life history analysis are indicated with vertical dashed lines.

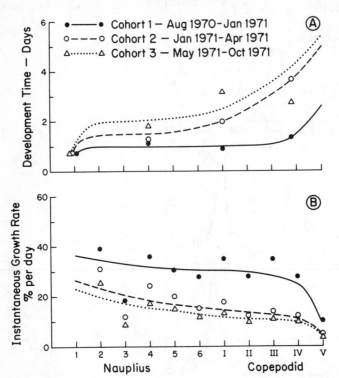

Figure 8–4. (A) Development time plotted against instar for calanoid copepods. Separate curves are drawn for three cohorts representing different portions of the study period. (B) Growth rate per day plotted against instar.

growth rates of the youngest instars are below those of the same cyclopoid instars, but growth rates of the later instars equal or exceed those of the cyclopoids.

Chaoborus

The development rates of the four instars of *Chaoborus* larvae were determined by a cohort analysis similar to that used for copepods. Figure 8–5 shows the seasonal patterns of abundance in the four instars and the three cohorts that were identified for analysis. It is clear from Figure 8–5 that the abundance patterns are much more irregular than for copepods. This is caused by large fluctuations from week to week in the egg-laying success of emerging adult *Chaoborus* (cf. McGowan 1974).

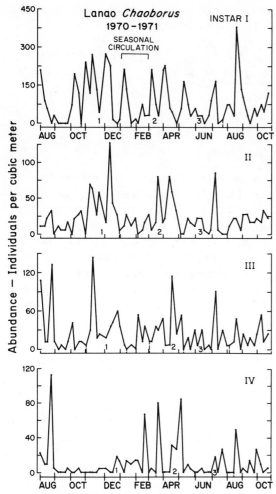

Figure 8–5.   Changes in abundance with time of developmental stages of *Chaoborus* in Lake Lanao. The three cohorts selected for life history analysis are indicated with vertical dashed lines.

The *Chaoborus* were measured individually, so it would have been possible to plot the abundance of 55 different size classes as a function of time. This is not a productive approach for cohort analysis, however, since the pace of development is quite rapid and the abundance patterns consequently vary radically from one week to the next. The population is instead split into instars based on size. The relationship between length and size for the various instars is discussed in detail elsewhere (Lewis, 1975).

The cohort analysis yields estimates of T_k, the temporal center of mass for kth instar. The method by which the duration of instars was obtained from the T_k values is slightly different than in the copepod cohort analysis because only four instars are involved. The approximations are as follows:

$$D_1 = T_2 - T_1$$

$$D_2 = \frac{D_1 + (T_3 - T_2)}{2}$$

$$D_3 = \frac{D_2 + (T_4 - T_3)}{2}$$

$$D_4 = T_4 - T_3$$

where D_k indicates the development time of instar k. If the duration of instars increases toward the terminal instar, this method will tend to overestimate D_1 and underestimate D_4. The data will show, however, that the instars do not differ greatly in duration, so the effect of the error is not great.

Figure 8–6 shows the development time for the four instars of the three cohorts identified in Figure 8–5. It was not possible to identify the fourth instar of cohort 1 with any certainty because of the low abundances of instar four during late 1970, so the fourth point is omitted for cohort 1. In general, the development times of the three cohorts are surprisingly similar and show

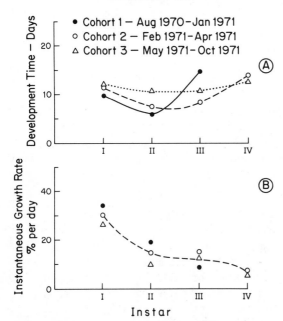

Figure 8–6. (A) Development time plotted against instar for *Chaoborus*. Separate curves are drawn for three cohorts representing different portions of the study period. (B) Growth rate per day plotted against instar.

similar trends. There is a tendency for the development time of the first and last instars to be slightly higher than for the intermediate instars.

The weight change in the course of each instar is known from the length–weight relationship for *Chaoborus* (Lewis, 1975). Given this relationship and the data in Figure 8–6, it is possible to approximate the instantaneous growth rate per day on the assumption that growth rate within instars is exponential. The result of this calculation is shown in Figure 8–6. The points for the three cohorts are so close together that a single curve will be used to represent all three cohorts. The curve shows that the relative growth of the first instar organisms is considerably more rapid than that of later instars.

Few studies are available which might serve as a basis for evaluating Figure 8–6. Studies done by Cressa (1971) on a tropical *Chaoborus* population suggest development times of approximately 30–45 days, and MacDonald's (1956) data for Lake Victoria are similar (47—62 days). The data of Comita (1972) suggest complete development of a temperate *Chaoborus* population over a period as short as 14 days. Unpublished studies by J. F. Saunders (personal communication) on Lake Valencia, Venezuela, suggest development times of only a few weeks. The development rates given in Figure 8–6 thus seem generally similar to development rates of other populations that have been studied.

Quite a number of laboratory feeding rate studies or field studies based on confinement have been done on *Chaoborus* (summarized by Lewis, 1977a). These studies all suggest food intake rates of less than 15% of body weight per day, and the average is only a few percent per day. As noted by Lewis (1977a) and Pastorok (1978), these percentages are low by comparison with other carnivores. Such feeding rates could not sustain the growth shown in Figure 8–6. Unfortunately, none of the students of feeding rates, including Lewis (1977a), has noted that the feeding rates determined in the lab are completely incompatible with observed development rates in field populations, either in temperate or tropical populations. For purpose of illustration, let the range of larval weights be about 4 to 3400 μg/individual (wet), as in Lanao. Most populations mature in 30 to 60 days of growing season (discounting overwintering), either in temperate or tropical lakes. Thirty-day maturation with exponential growth would require growth rates of about 25% per day, and 60-day maturation would require growth rates of about 12% per day. Even with high assimilation rates and growth efficiencies, intake must be four to five times as high as growth. *Chaoborus* populations must therefore routinely ingest amounts of prey ranging from 50% to more than 100% per day of the total population weight, much more than suggested by laboratory studies. Confinement apparently reduces intake rates and thus produces very misleading estimates of intake. Part of the discrepancy may be explained by the focus of most investigations on the last instar, which has the lowest growth rate (Fig. 8–6), but suppression of normal feeding behavior is almost certainly involved as well. One can only hope that the extensive laboratory studies of electivity are not similarly biased. The computations of the present analysis will rely entirely on field measurements not involving confinement.

Rotifers and Cladocera

Development times for rotifer eggs were not studied in the laboratory. Sufficient numbers of development times have been measured and reported in the literature to allow a reasonably confident prediction of development time for eggs in the Lanao populations (Table 8–1). The values in the table have been corrected to 25°C, the temperature of Lake Lanao, by use of Krogh's curve (Winberg, 1971).

It is also possible to approximate the development time for eggs from field data given certain assumptions, although such an approach has apparently never been reported in the literature. The procedure is as follows.

1. For each rotifer species, the change in population size (N_2-N_1) is computed for each week in the study.

2. The assumption is made that growth for any 1-week period is exponential. Thus the rate of increase each week is computed as follows:

$$r = \frac{\ln(N_2) - \ln(N_1)}{t_2 - t_1}$$

where r is the instantaneous rate of increase in the population and N_k is the population size at time t_k. It will become evident that this assumption need be true only when the population is growing quite rapidly for the method of computation to be valid.

3. On occasions when the population is growing very rapidly, the instantaneous growth rate (b) will be considerably larger than the instantaneous death rate (d). These periods of rapid increase give us a means of estimating b, since $r = b - d$ and $b \gg d$, so that $r \simeq b$.

4. The development time for eggs (D) is computed from the egg ratio (E) and b as follows.

$$b = \ln(B + 1)$$
$$B = e^b - 1$$
$$B = E/D$$
$$D = E/(e^b - 1).$$

The conversion of finite birth rate (B) to an instantaneous rate (b), as shown here, actually involves a certain degree of error because of mortality over the estimation interval (Caswell, 1972; Edmondson, 1972). Alternatives have been proposed (Edmondson, 1968; Caswell, 1972; Paloheimo, 1974), but these will probably not be superior under all con-

Table 8–1. Egg Development Times and Mean Egg Ratios for Rotifers of Lake Lanao[a]

Species	Egg development, days —25°C			Egg ratios (eggs/female)		
	From literature	From field data	Used in computation	Mean	n	Coefficient of variation $(s/\bar{x}) \cdot 100$
Conochiloides dossuarius	1.2[b]	1.2	1.2	0.28	49	34
Hexarthra intermedia	—	1.2	1.2	—	—	—
Polyarthra vulgaris	0.9[c]	—	1.0	—	—	81
Keratella procurva	—	—	1.0	0.26	5	53
Keratella cochlearis	0.9[c]	1.2	1.1	0.18	25	—
Trichocerca brachyurum	—	—	1.2	—	—	—
Tetramastix opoliensis	—	1.0	1.0	0.12	23	51

[a] Development times from the literature are corrected to 25°C and are compared with values computed from field data by the method described in the text. Egg ratios were computed over all dates meeting statistical criteria specified in the text.
[b] Makarewicz (1975).
[c] Edmondson (1965).

ditions. These points are of minor concern here because the method focuses on time periods when births greatly exceed deaths.

The most serious pitfall connected with this method of estimates occurs in the selection of periods when the value of r is high. The very lowest rotifer abundances have a high error variance because of the counting procedures on which the abundance estimates are based. Thus these estimates may result in spuriously high values of r, which must be screened out prior to the time any computations are made. For the Lanao data, the number of organisms counted each week is known (Table 4–1), hence an estimate can be made of the error variance to be expected for the abundance estimates each week. The criterion is adopted that abundance estimates for days on which the error variance attributable to counting would generate a coefficient of variation $[(s/\bar{x}) \cdot 100]$ greater than 10% should not be included in the analysis. The b values remaining after this screening process are tabulated in descending order and the top 10% of these are selected for use in the computation of D.

The development times for eggs obtained from the field data as outlined above are reported in Table 8–1. It was not possible to make estimates for some of the rarer species owing to the small number of dates on which these species reached detectable abundances. The numbers reported in Table 8-1 are in good agreement with development times obtained from the literature. This inspires confidence in the method and in the development times to be used in further analysis.

Development times for Cladocera eggs were obtained from field data by a procedure identical to that used for rotifers. The development times are reported in Table 8-2.

Surprisingly little information exists on the development times of non-daphnid Cladocera eggs. Keen (1973) has shown that the development time for eggs of four littoral species of Cladocera is about 1.7 days at 25°C. The literature review of Bottrell et al. (1976, Fig. 11) suggests that the mean development time at 25°C for eggs of Cladocera other than *Daphnia* will be slightly in excess of 1 day. These figures are in good agreement with Table 8-2.

Table 8–2.   Egg Development Times as Determined from Field Data and Mean Egg Ratios for Cladocerans in Lake Lanao

Species	Egg development, days—25°C	Egg ratios (eggs/female)		
		Mean	n	Coefficient of variation $(s/\bar{x}) \cdot 100$
Diaphanosoma modigliani	1.50	0.31	37	50
Moina micrura	2.15	0.53	3	63
Bosmina fatalis	1.42	0.22	9	56

Chapter 9

Secondary Production

Computations: Copepods and *Chaoborus*

For calanoid and cyclopoid copepods and for *Chaoborus* the production of each developmental stage was computed separately as follows (Edmondson, 1971; Winberg, 1971):

$$P_i = \frac{N_i \, \Delta w_i}{D_i}$$

where P_i is the production of stage i, N_i is the numerical abundance of stage i, Δw_i is the change in weight over stage i, and D_i is the development time for stage i. The values of D_i were obtained from the cohort analysis, which has already been discussed. The values of P_i were summed as required to obtain the production of groups of developmental stages or of the entire population.

Computations: Rotifers

Rotifers lack the multiple developmental stages of copepods and *Chaoborus* and develop to maturity very rapidly. A cohort analysis followed by summation of production over developmental stages is therefore impossible for rotifers. Computations of rotifer production are usually based on estimates of the birth rate obtained by the egg ratio method of Edmondson (Edmondson, 1968, 1971). This method, which modifies the method of Elster (1954, 1955) by using exponential rather than linear equations, allows the computations to reflect changes in the field population because the egg ratio (eggs per female) which is used in the computation of birth rate is taken from field populations. Although an estimate is also required of egg development rate, egg development is principally sensitive to temperature and thus can be predicted without continuous field measurements, especially since rotifers are not so inclined as copepods and cladocerans to migrate across a broad temperature range.

The egg ratio technique as generally applied for production estimates poses some difficulties in connection with the growth which occurs between the egg and the adult. This difference will be referred to here as "postinduction growth," meaning the increase in size from the beginning of cleavage to maturity. The term "postembryonic growth" is sometimes used in this context but is less definite because of the uncertainty of the termination of the embryonic phase.

Postinduction growth is typically ignored in application of the egg ratio method. The postinduction growth of rotifers is much lower than that of other major zooplankton categories, but the change of size in rotifers after hatching is by no means trivial (Table 4-1). Although the cell number of individuals does not change after cleavage is complete (Ruttner-Kolisko, 1974), the increase in size of individual cells can be substantial and of course contributes to the growth in biomass of the population.

As an alternative to the egg ratio method for rotifers some authors (e.g., Makarewicz, 1974) have used the method of Galkovskaya (1965, as cited in Winberg, 1971) based on generation time, which is actually very similar to the method used by Stross et al. (1961) (Edmondson, 1971, p. 311). The computation is:

$$P = B/G$$

where P is daily production, B is the total biomass of the population, and G is the generation time in days. The generation time method avoids the assumption that postinduction growth is negligible. The difficulty is that the generation time (G) is typically determined in the laboratory or at least under controlled conditions of some kind. The final estimate is therefore not apt to reflect the true changes in vigor of the field populations. While laboratory estimates seem defensible for egg development on grounds that eggs do not feed and are thus mainly affected by temperature, laboratory estimates for egg synthesis and postinduction growth cannot be defended as easily because the growing animals will be affected in very unpredictable ways by the quality and quantity of their food resource.

In summary, the egg ratio method seems unsatisfactory because it assumes negligible postinduction growth, and the generation time method seems unsatisfactory because it assumes that laboratory estimates of generation time will reflect field conditions. Concepts from both methods are useful, but it would be desirable to combine the virtues of the two methods.

Such a combination is actually quite feasible. Consider the equation for production of a single developmental stage of one of the copepods according to the standard method used earlier on copepods:

$$P_i = \frac{N_i \, \Delta w_i}{D_i} \, .$$

If the organism in question (a rotifer) can be considered to consist of a single developmental stage ($i = 1$) incorporating synthesis of egg protoplasm plus postinduction growth, then D_i is the generation time, N_i is the total number of individuals (eggs + adults), and Δw_i is the total pre- and postinduction growth. Recognizing that only one stage is involved, we have:

$$P = \frac{N \, W_a}{G}$$

where W_a is the weight of a fully grown animal. Production will be computed from this equation. The egg ratio will be used to estimate the generation time of field populations and the generation time for a given week will then be combined with the abundance of the population and adult weight as shown in the equation. The implementation of this method will require (1) the egg ratio, (2) the development time for eggs (Chapter 8), and (3) a rationale for obtaining generation times from egg ratios.

The determination of egg ratios from field data is subject to a number of errors (Edmondson, 1971). Escape of adults or eggs through the meshes of nets (Likens and Gilbert, 1970; Doohan, 1973; Bottrell et al., 1976) and rhythmicity in oviposition (Edmondson, 1965) are the two most serious errors. Escape of organisms through the nets was not a problem in the present study (see Chapter 5), so only rhythmicity in oviposition need be considered here.

The time of zooplankton sampling alternated between mid-morning and evening throughout most of the study. The alternating samples provide a means for detecting rhythmicity in the population. If the egg ratio is not significantly influenced by time of day, then the average egg ratio for night samples will be the same as the average egg ratio for day samples. Night samples and day samples for all species were compared statistically and no differences were found at the 5% level (Sign test, Sokal and Rohlf, 1969). For *Conochiloides,* the most common rotifer species, the mean egg ratio for night samples is 0.26 whereas the mean egg ratio for day samples is 0.30. Although the data fail to show a pronounced rhythmicity in oviposition, they do not rule out the possibility of an irregular diel variation in egg ratios or a rhythmicity that changes with the seasons. Such irregular variations would not be as strong a source of bias in the final production estimates, however.

The next problem is to obtain an estimate of the generation times of field populations from their egg ratios. For an individual organism, the generation time based on number of individuals (G_i) and the generation time based on biomass (G_b) are both equal to the amount of time required to pass completely through the life cycle one time. For a population growing exponentially, the generation time based on individuals is:

$$G_i = \frac{\ln(2)}{b}.$$

The variable b is the instantaneous birth rate and can be approximated from the egg ratio (E) and egg development time (D) by the equation:

$$b = \ln(E/D + 1).$$

Assumptions inherent in the conversion of finite to instantaneous rates and in the use of the exponential equation to predict the behavior of a population over a short interval have been dealt with at length by Edmondson (1971).

For a population, the variable G_i in the above equation is equal to G_b if the age structure of the population is unchanging. G_i may differ from G_b, however, if the age distribution of individuals is not constant. For rotifer populations, it is convenient to recognize two separate problems in this connection. The first of these derives from the possibility that the synthesis of protoplasm incorporated in the egg may not occur at the same rate as the synthesis of protoplasm during postinduction growth. The second problem is that the egg does not increase in biomass while it is developing, so that the presence of a large number of eggs may actually decrease the turnover time of biomass.

No published information exists on the first problem, so it can be dealt with only by the assumption that pre- and postinduction growth rates are similar. The second problem severely biases the estimate only if the ages of eggs are not randomly distributed at a given time. Since evidence has already been presented which suggests that no persistent rhythmicity exists in the production of eggs by the Lanao populations, the assumption of randomly distributed egg ages will be acceptable unless some kind of strong seasonally alternating rhythmicity exists, which seems unlikely. For these reasons, G_i will be considered equal to G_b so that either of these is equal to the generation time (G) in the production equation.

Since the generation times are computed for each week from the egg ratios and the foregoing equations, the generation times will differ from week to week. On some weeks, however, reliable measurement of the egg ratio is not possible owing to low abundance of a species. Particular care must be taken to rule out time periods for which the counts are not reliable, as the incorporation of two different abundance estimates (eggs and females) in a ratio will magnify counting error. This difficulty is of course not unique to the present method but arises whenever statistics are taken from field populations. To reduce distortion, a statistical screening criterion identical to that which has already been described for computation of D is carried through to the production calculations. Egg ratios are computed for those time periods represented by counts which have a coefficient variation of less than 10%. When it is not possible to compute an egg ratio for a particular period, that period will be considered to have an egg ratio equal to the mean egg ratio on all periods for which computations were possible.

Table 8–1 summarizes the mean egg ratios for time periods meeting the statistical criterion for each of the species. The amount of variation in the egg ratios over the course of the study period is surprisingly low. For example, the egg ratios of *Conochiloides,* the most common rotifer species, all fall between 0.20 and 0.47, with one exception. The egg ratios of temperate populations that have been studied appear to vary considerably more (Edmondson, 1965).

Computations: Cladocera

Cladoceran populations would ideally be divided into size classes and the production would be figured by methods similar to those used for copepods. Cladocerans are a more minor element of the zooplankton in Lake Lanao than copepods and were not individually measured, so they cannot be divided into size classes. Production computations are therefore done by the same method as used for rotifers. The method is less ideal for Cladocera than for rotifers because of the larger amount of postinduction growth which occurs in the Cladocera. The larger size range of individuals makes differential growth and mortality much more probable than in rotifers. The cladoceran species of Lake Lanao are extremely small (Table 4–1), however, which to some extent minimizes these difficulties. *Bosmina* in particular grows so little between the egg and age of first reproduction that it is more similar to rotifers than to copepods in this respect (Kerfoot, 1974).

The cladoceran populations were checked for rhythmicity in oviposition by a technique identical to that used for rotifers. No significant differences between day and night samples exist for the egg ratio of any of the three species. For *Diaphanosoma,* the most common of the cladoceran species, the egg ratio in night samples is 0.29 and for day samples it is 0.36. Since this difference is not statistically significant, it will be assumed that no persistent rhythmicity exists in the laying of eggs. The mean egg ratios for each species are shown in Table 8–2.

Results: Copepods and *Chaoborus*

Table 9–1 summarizes the total production of *Thermocyclops* and the relative contribution of all of its developmental stages. The productivities of individual stages, with the exception of Nl and CV, are surprisingly equal in

Table 9–1. Production of Zooplankton Species and Developmental Stages in Lake Lanao over the Study Period (Wet Weight)[a]

Species/stage	Production (μg/liter/day)	Coefficient of variation ($s/\bar{x} \cdot 100$)	Percentage of species production	Percentage of class production (copepod, Cladocera, rotifer)	Percentage of total production
Copepods					
Thermocyclops hyalinus					
Nauplius 1	0.73	129	3.88	2.71	1.79
2	1.28	73	6.81	4.75	3.14
3	1.73	65	9.20	6.43	4.25
4	1.62	91	8.62	6.02	3.98
5	1.63	159	8.67	6.05	4.00
6	2.50	149	13.30	9.29	6.14
Σ	9.49	71	50.48	35.25	23.30
Copepodid I	2.72	84	14.47	10.10	6.68
II	2.25	84	11.97	8.36	5.52
III	1.81	81	9.63	6.72	4.44
IV	1.88	73	10.00	6.98	4.62
V	0.21	73	1.12	0.78	0.52
Σ	8.86	71	47.13	32.91	21.75
Adult (=egg)	0.51	103	2.71	1.89	1.25
All stages	18.80	73		69.84	46.16
Tropodiaptomus gigantoviger					
Nauplius 1	0.06	93	0.74	0.22	0.15
2	0.10	89	1.23	0.37	0.25
3	0.15	78	1.85	0.56	0.37
4	0.44	82	5.42	1.63	1.08
5	0.31	76	3.82	1.15	0.76
6	0.38	97	4.68	1.41	0.93
Σ	1.44	73	17.73	5.35	3.54

Copepodid I	0.94	98	11.58	3.49	2.31
II	0.74	84	9.11	2.75	1.82
III	1.74	87	21.43	6.46	4.27
IV	1.03	66	12.68	3.83	2.53
V	1.50	81	18.47	5.57	3.68
Σ	5.94	62	73.15	22.07	14.58
Adult (=egg)	0.74	76	9.11	2.75	1.82
All stages	8.12	56		30.16	19.94
All copepods	26.92	60			66.10
Cladocera					
Diaphanosoma modigliani	4.84	110	100.0	76.58	11.88
Moina micrura	0.78	213	100.0	12.34	1.92
Bosmina fatalis	0.79	293	100.0	12.50	1.94
All Cladocera	6.32	113			15.52
Rotifers					
Conochiloides dossuarius	0.79	144	100.0	68.10	1.94
Hexarthra intermedia	0.14	205	100.0	12.07	0.34
Polyarthra vulgaris	0.04	272	100.0	3.45	0.10
Keratella procurva	0.02	205	100.0	1.72	0.05
Keratella cochlearis	0.06	178	100.0	5.17	0.15
Trichocerca brachyurum	0.03	243	100.0	2.59	0.07
Tetramastix opoliensis	0.08	145	100.0	6.90	0.20
Total rotifers	1.16	109			2.85
Diptera					
Chaoborus instar I	0.56	95	8.85	8.85	1.37
II	0.65	96	10.27	10.27	1.60
III	2.49	112	39.34	39.34	6.11
IV	2.62	171	41.39	41.39	6.43
Total	6.33	104			15.54
Total herbivores	34.40	57			84.46
Total carnivores	6.33	104			15.54

[a] To convert $\mu g/liter/day$ to $mg/m^2/day$, multiply by 45.

magnitude. The total production in the naupliar stages is about the same as the total production in the copepodid stages. Egg production is a very small proportion of total production for the species. This is not atypical for copepods. Rigler and Cooley (1974), working with Eichhorn's (1957) data on Titisee (*Mixodiaptomus lacinatus*), showed that egg production accounted for only 3.5% of total production for the species. The same authors showed a comparable figure of 7.0% from Elster's (1936) data (*Heterocope borealis*) and about 10% for their own data (*Skistodiaptomus oregonensis*).

Figure 9–1 shows the seasonal production patterns for *Thermocyclops*.

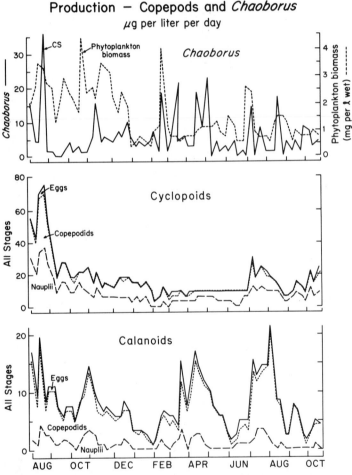

Production — Copepods and *Chaoborus*
μg per liter per day

Figure 9–1.   Production as a function of time for cyclopoid copepods, calanoid copepods, and *Chaoborus* (wet weight). Phytoplankton biomass in the euphotic zone is also shown for comparative purposes. The areas under the curves for both cyclopoid and calanoid copepods have been divided to show the relative contribution of nauplii, copepodids, and eggs to the total production of the population.

Although there are considerable surges in production, a significant amount of production is sustained throughout the year.

Tropodiaptomus is numerically less abundant than *Thermocyclops* and accounts for a smaller proportion of the total production (Table 9–1). The production is distributed much less equally between developmental stages in *Tropodiaptomus* than in *Thermocyclops*. Naupliar stages account for considerably less of the total production than copepodid stages. Egg production is a more important component of total production than for *Thermocyclops*, but is still only a small proportion of total production for the species. Total variation in production is less than for *Thermocyclops* and does not show such pronounced seasonality (Fig. 9–1).

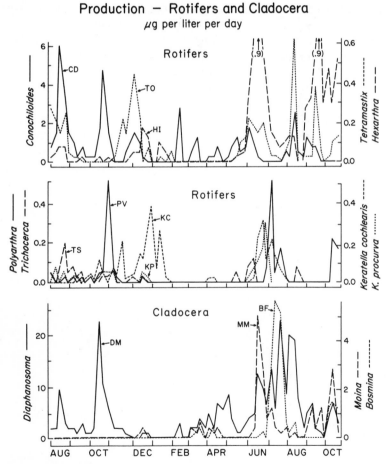

Production — Rotifers and Cladocera
µg per liter per day

Figure 9–2.   Production as a function of time for rotifers and Cladocera in Lake Lanao (wet weight). The two-letter codes are derived from the first letter of the genus name and the first letter of the species name.

The production of *Chaoborus* is also shown in Figure 9–1 as a function of time. By comparison with copepod production, *Chaoborus* production is more irregular and shows no evidence of seasonality. Table 9–1 shows that the bulk of production occurs in the larger instars. *Chaoborus* production, which is nourished entirely from the herbivore production, is 18% of total herbivore production.

Results: Rotifers and Cladocera

Rotifers account for a very small percentage of the total secondary production in Lake Lanao (Table 9–1). *Conochiloides dossuarius* is the dominant species in terms of production. There is a pronounced seasonal pattern in the production of all species, but the seasonal pattern is distinct from patterns observed for Cladocera and copepods (Fig. 9–2).

The Cladocera account for a significant percentage of total production but are much less important than copepods in this respect (Table 9–1). *Diaphanosoma* is by far the dominant cladoceran in terms of production. For *Diaphanosoma* the minima in production are similar in timing to those observed for copepod populations, but details of variation are considerably different. Variations in populations of *Moina* and *Bosmina* are even more irregular than for *Diaphanosoma* (Fig. 9–2).

Chapter 10

Similarities in Zooplankton Abundances and Productivities

The seasonal patterns in abundance and productivity of individual species are too complex to permit much generalization concerning the differences and similarities between species. A more rigorous approach is possible using correlation matrices derived from the variation in population properties with time. Four different comparisons are possible with the population statistics that have been developed up to this point.

1. Populations can be compared with respect to their variation in abundance through time. In terms of a correlation matrix, it is inconsequential whether abundance is measured as numbers or biomass. In the present analysis, interpretation is based on the correlation matrix for numerical abundance through time for all species, which will be referred to as the N matrix. The N matrix integrates the past history of a population over an indefinite time period. This is illustrated in Figure 10–1. For populations with very low turnover, the effective integration extends back further in time than for populations with very high turnover. This inhibits the usefulness of the N matrix for any analysis which attempts to compare populations over successions of equal time intervals. Thus the principal value of the N matrix is in (a) description of community composition, and (b) analysis of the use of zooplankton populations as a resource (e.g., by predators).

2. It is also possible to compute for each species the change in abundance (ΔN) for each week in the study period. This gives rise to the ΔN matrix. The ΔN matrix is based on an approximation of the first derivative of abundance with respect to time, and thus requires a completely dif-

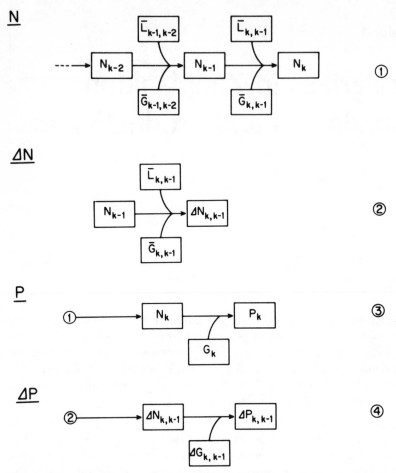

Figure 10–1. Diagrammatic illustration of the factors affecting variables used in the discussion of similarities between species (N, ΔN, P, ΔP). Symbols are as follows: N, abundance; P, production; G, growth; L, loss; k, time interval.

ferent interpretation from the N matrix. As indicated elsewhere in connection with the analysis of phytoplankton communities (Lewis, 1977b), the ΔN matrix has considerable analytical power with regard to the ecological similarities between species.

The magnitude of ΔN over a fixed time interval is a composite measure of the response of the species to the entire complex of environmental effects, including both growth control and loss control factors (Fig. 10–1). Thus over a long succession of time intervals the similarity in values of ΔN is a measure of overall ecological similarity of species or stages within a given trophic level. Obviously an interpretation of predator–prey correlations (i.e., *Chaoborus* with any of the other species) must be made on a different basis, however.

3. It is also possible to compute a matrix of correlation coefficients based on the production of individual species through time (*P* matrix). This *P* matrix is, like the ΔN matrix, based on a first derivative of abundance with respect to time. The ΔN matrix, however, integrates the effects of attrition and growth over the course of a time interval, whereas the *P* matrix is an instantaneous estimate of the synthesis of new individuals or new protoplasm, including that which is destined to be eaten or lost to other sources of attrition (Fig. 10–1). A population may sustain high production without increasing its numbers or biomass and the value of ΔN can thus be 0 even though the value of *P* is high. From an analytical viewpoint, ΔN is more revealing with respect to overall ecological similarity, as it provides an estimate of the success of a species in increasing its abundance in the environment, whereas *P* is more useful in evaluating similarity of species with regard to growth control mechanisms, including any feedback effect of mortality on growth and the secondary effect of loss rates on *N*, but not loss control factors directly. Moreover, the *P* matrix, because it is strongly dependent on *N*, has a strong historical component as does *N* itself and thus is not so well suited as ΔN for comparisons between populations with different turnover rates. The *P* matrix is ideal, however, as an indication of similarities in demands on food resources within a given trophic level.

4. Similarities between species in change of productivity (ΔP) from one week to the next can also be expressed in terms of a correlation matrix (ΔP matrix). Since *P* itself is based on a first derivative of biomass with respect to time, ΔP is based on a second derivative of biomass with respect to time. ΔP is more similar in its analytical implications to ΔN than it is to *P*. Since ΔP over a fixed time interval integrates the response of a population to conditions over that time interval, it provides a good means of comparing the overall ecological similarity of species. The critical difference from ΔN is that ΔP, like *P* itself, is more directly influenced by the mechanisms that control growth than by the mechanisms that control mortality, whereas ΔN is unbiased in this respect (Fig. 10–1). The value of ΔP is of course not entirely independent of factors affecting attrition, since changes in population density or age structure have a feedback effect on *P*. The linkage between ΔP and factors which control mortality is nevertheless more indirect than between ΔN and factors which affect mortality. This means that the interpretation of the ΔN and ΔP matrices must be made separately.

In summary, overall ecological similarity between species is best measured with the ΔN matrix, but the *N* matrix is still useful for indicating the co-occurrence of populations in time or the coincidence of their availability to a predator. The ΔP matrix allows determination of ecological similarity based more strictly on growth control mechanisms. The P matrix is useful for identifying the coincidence in demand that species place on the resource pool, but less so for identification of ecological similarities.

The correlation matrices of N, ΔN, P, and ΔP values are shown in Tables 10–1 and 10–2. All species are represented in the matrix, including the predator *Chaoborus*. The two copepod species are broken down into three separate stages (naupliar, copepodid, and adult) on grounds that the tremendous change in size and morphology of these species in the course of development may allow them to overlap with different herbivore species at different phases in their life cycle.

Some of the species pass through great periods of rarity. At such times it is not possible to estimate the abundance of the species, because it may not appear in even the largest sample. For the sampling and counting methods that were employed on Lake Lanao, one would expect to see an average of about five individuals of a species whose abundance in a weekly sample series is 0.10 individuals per liter. The data were screened so that all dates on which the abundance of a species fell below 0.10 individuals per liter were excluded from the computation of a correlation coefficient for N, ΔN, P, or ΔP values. For the rarest rotifers, this resulted in exclusion of approximately half of the data points. Obviously this screening is necessary in order to avoid the production of spuriously significant correlation values which would be produced as the byproduct of a substantial number of 0 values in the raw data matrices.

The amount of autocorrelation in the raw data matrices must be considered insofar as a large amount of autocorrelation can produce spuriously high correlation values (Denman, 1975; Winkler and Hays, 1975). The matrices of N, ΔN, P, and ΔP values were checked for autocorrelation with time lags of 1, 2, 3, 4, and 5 weeks. Not all species and stages could be included in the autocorrelation analysis because of the number of 0 values in the abundance matrices for some species. The analysis was consequently limited to the six copepod categories shown in Table 10–1 and *Conochiloides,* the dominant rotifer.

The N and P matrices showed the greatest amount of autocorrelation. For the N matrix, 14 of the 35 autocorrelation coefficients (5 lags $\times$ 7 species or stages) were significant at the 0.05 level. Almost all of these significant values were produced as a result of 1- or 2-week lags. The results for the P matrix were almost identical. This is to be expected in view of the large influence N has on the estimate of P. These results suggest that, while autocorrelation is limited primarily to lags of 1 or 2 weeks, correlation coefficients for the N and P matrices should be interpreted more conservatively than if the data were not autocorrelated at all. For the ΔN and ΔP matrices there is virtually no autocorrelation. Of the 70 correlation coefficients that were com-

Table 10–1. Matrix of Pearson Product–Moment Correlation Coefficients for N (above the Diagonal) and for ΔN (below the Diagonal)[a]

	Copepods						Rotifers							Cladocera			Chaoborus
	CYN	CYC	CYA	CAN	CAC	CAA	CD	KC	TO	KP	HI	PV	TS	DM	MM	BF	CS
CYN	+	0.49*	0.54*	0.55*	0.44*	0.37*	0.41*	−0.20	0.06	0.11	−0.22	0.06	0.47*	0.49*	−0.07	0.54*	0.19
CYC	0.51*	+	0.69*	0.22	0.00	0.24	0.61*	0.14	0.24	0.28	−0.08	0.26	0.70*	0.26*	−0.02	0.46*	−0.01
CYA	0.23	0.46*	+	0.31*	0.06	0.44*	0.41*	−0.02	0.23	0.29	0.06	0.14	0.49*	0.29*	0.21	0.24	0.21
CAN	0.38*	0.27*	0.07	+	0.53*	0.62*	0.15	−0.15	−0.11	0.10	−0.09	0.40	0.34	0.40*	−0.08	0.71*	0.13
CAC	0.36*	0.35*	0.10	0.28*	+	0.21	−0.02	−0.27	0.01	−0.21	−0.03	0.39	−0.08	0.67*	−0.18	0.39*	−0.07
CAA	0.24	0.23	0.36*	0.43*	0.23	+	0.11	−0.17	0.04	0.29	0.00	0.30	0.25	0.14	−0.02	0.04	0.22
CD	0.25*	0.31*	0.02	0.30*	−0.06	−0.05	+	−0.36*	0.48*	0.47*	−0.07	−0.19	0.83*	0.31*	−0.20	−0.20	−0.09
KC	0.00	0.05	0.14	0.00	0.12	−0.08	−0.10	+	−0.04	−0.02	−0.07	0.35	−0.22	−0.05	0.54*	0.39	0.01
TO	0.06	0.10	0.07	−0.03	−0.18	−0.03	0.43*	0.03	+	0.20	0.25	0.15	0.31	0.31*	−0.07	−0.12	−0.02
KP	0.01	0.06	−0.06	0.20	−0.05	0.03	0.37*	−0.39*	−0.28	+	0.25	−0.19	0.53*	0.28	0.66*	−0.41	−0.19
HI	−0.02	0.04	0.12	−0.12	−0.01	0.07	0.08	−0.01	0.23	0.39	+	0.25	−0.18	0.01	−0.42*	−0.24	−0.19
PV	0.31	0.58*	0.23	0.18	0.54*	0.49*	−0.02	0.31	−0.15	−0.33	−0.02	+	−0.22	0.36	0.20	0.19	0.20
TS	0.01	0.06	−0.16	0.43*	0.16	0.05	0.50*	−0.09	0.17	0.51*	−0.08	0.05	+	0.26	−0.26	−0.30	−0.11
DM	0.42*	0.35*	0.09	0.28*	0.50*	0.18	0.24	0.04	0.11	0.07	0.09	0.30	−0.02	+	0.05	0.27	−0.03
MM	0.13	−0.20	0.04	0.13	0.13	0.18	−0.06	0.36	−0.22	0.03	0.13	0.36	−0.48	0.17	+	0.01	−0.20
BF	0.16	−0.16	−0.15	0.64*	−0.08	−0.07	−0.07	−0.01	−0.06	−0.10	−0.12	−0.11	−0.06	−0.09	0.16	+	−0.17
CS	0.57*	0.45*	0.37*	−0.08	0.26	0.20	−0.03	0.03	0.05	−0.35	−0.02	0.53*	−0.11	0.20	−0.06	−0.35	+

[a] Coefficients that are significant at $P = 0.05$ are marked with an asterisk.
[b] Letter codes are from genus and species names except for cyclopoids (CY) and calanoids (CA) with their developmental stages (nauplius, copepoid, adult).

Table 10–2. Matrix of Pearson Product–Moment Correlation Coefficients for P (above the Diagonal and for ΔP (below the Diagonal)[a]

	Copepods							Rotifers						Cladocera			Chaoborus
	CYN	CYC	CYA	CAN	CAC	CAA	CD	KC	TO	KP	HI	PV	TS	DM	MM	BF	CS
CYN	+	0.84*	0.68*	0.50*	0.37*	0.33*	0.54*	-0.16	0.14	-0.11	-0.16	-0.07	0.66*	0.17	-0.09	0.66*	0.35*
CYC	0.67*	+	0.70*	0.47*	0.20	0.28*	0.55*	0.18	0.13	-0.14	-0.19	0.17	0.70*	0.05	-0.01	0.59*	0.30
CYA	0.31*	0.46*	+	0.44*	0.19	0.47*	0.33*	0.04	0.19	0.20	-0.03	-0.08	0.49*	0.09	0.27	0.30	0.49*
CAN	0.30*	0.39*	0.04	+	0.53*	0.61*	0.41*	-0.06	-0.14	-0.06	-0.23	0.55*	0.55*	0.30*	-0.09	0.61*	0.20
CAC	0.29*	0.35*	-0.02	0.40*	+	0.29*	0.19	-0.21	-0.05	-0.35*	-0.20	0.52*	0.27	0.48*	-0.23	0.22	0.09
CAA	0.28*	0.22	0.38*	0.35*	0.16	+	0.11	-0.10	0.00	0.21	-0.12	0.11	0.32	0.04	0.00	0.08	0.38*
CD	0.28*	0.30*	-0.02	0.40*	0.04	-0.08	+	-0.28	0.30*	-0.11	-0.12	-0.07	-0.73*	0.22	-0.30	-0.26	0.04
KC	-0.13	0.01	0.05	-0.01	0.11	-0.17	-0.06	+	0.05	0.26	0.22	0.18	-0.24	0.10	0.68*	0.32	0.05
TO	0.05	-0.04	0.08	-0.11	-0.26*	-0.06	0.30*	0.16	+	0.12	0.35*	-0.10	0.13	0.12	0.02	-0.11	0.05
KP	-0.34*	-0.44*	0.12	-0.27	-0.56*	0.07	-0.13	-0.23	0.11	+	0.55*	-0.17	-0.10	0.10	0.41	-0.37	-0.03
HI	0.19	0.02	0.01	0.03	0.03	0.01	0.09	0.10	0.44*	0.35*	+	-0.14	-0.12	-0.01	0.34	-0.26	-0.03
PV	0.18	0.48*	0.07	0.48*	0.49*	0.14	-0.16	-0.01	-0.15	-0.43	-0.30	+	-0.13	0.26	-0.02	0.70*	-0.06
TS	0.02	0.09	-0.10	0.45*	0.37*	0.05	0.39*	-0.24	-0.17	-0.10	-0.02	0.08	+	0.05	-0.23	-0.23	0.14
DM	0.34*	0.20	-0.03	0.26*	0.32*	0.17	0.24	0.19	0.08	-0.22	-0.01	0.02	-0.07	+	0.22	0.44*	-0.03
MM	-0.03	0.02	0.21	0.12	0.15	0.21	-0.12	0.34	-0.07	-0.17	0.06	0.24	-0.64*	0.36	+	-0.04	-0.02
BF	0.34	0.20	-0.13	-0.54*	0.04	0.08	-0.04	0.18	0.01	-0.36	-0.06	0.46	-0.11	0.33	0.04	+	0.09
CS	0.54*	0.46*	0.59*	-0.04	0.10	0.43*	-0.16	0.09	0.10	-0.11	-0.02	0.18	-0.24	0.16	0.10	0.06	+

[a] Coefficients that are significant at $P = 0.05$ are marked with an asterisk.
[b] Letter codes are from genus and species names except for cyclopoids (CY) and calenoids (CA) with their developmental stages (nauplius, copepodid, adult).

puted for these two matrices, only three were signifcant at the 5% level and this number would be expected as a result of chance alone assuming no auto-correlation at all.

All four of the correlation matrices were subjected to cluster analysis to facilitate a discussion of similarities between species. The cluster analysis was in all cases based on the Pearson Product–Moment Correlation Coefficients as reported in Tables 10–1 and 10–2. Results are shown in Figure 10–2.

The frequency distributions for the variables N, ΔN, P, ΔP are not always normal. Tests of the frequency distributions for all species and for naupliar,

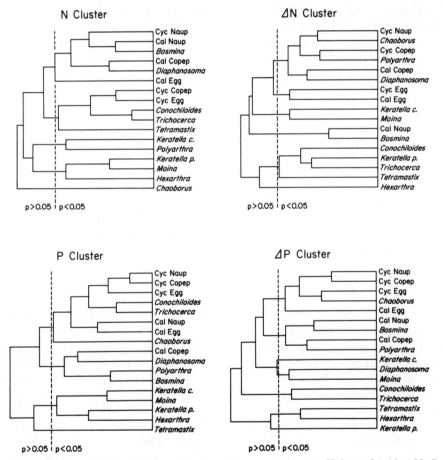

Figure 10–2. Cluster diagrams based on the correlation coefficients for N, ΔN, P, ΔP. The vertical dotted line approximates the 5% significance level for pairings, but the exact position of the 5% level will vary somewhat according to the identity of the variables being paired.

copepodid, and egg stages of copepods showed that the N and P values for a given species or stage are always distributed with a strong positive skew. This is also typical of phytoplankton populations (Lewis, 1977c). The ΔN and ΔP values are distributed symmetrically but with strong leptokurtosis. In general, the P and ΔP values differ more from normality than the N and ΔN values. Across species, distributions of all variables for copepods, *Tetramastix, Diaphanosoma,* and *Chaoborus* show the least departure from normality. Departure of the frequency distributions from normality may affect the correlation coefficients and thus is of some concern with regard to the analysis of the correlation matrices. For purposes of comparison, the matrices were all recalculated using the Spearman Rank Correlation Coefficient (Conover, 1971), which makes no assumptions about the normality of the data. The pattern of significant correlations in the matrices of Spearman Rank Coefficients was essentially the same as the pattern in the matrices of Pearson Product–Moment Coefficients. The Pearson Product–Moment Coefficients were therefore used in the cluster analysis and in the final interpretation of data.

The N Matrix

The N Matrix contains 128 unique pairings of organisms or stages, of which 35 (27%) produce significant correlations. With one exception, all of these statistically significant coefficients are positive. Since the significance level is 0.05, only six significant correlations should appear in the matrix as a result of chance alone, or perhaps slightly more than this number if autocorrelation is responsible for producing some spurious incidences of statistical significance. Thus most of the significant correlations are likely to underlie some actual biological relationship between the abundances of the organisms involved.

It is clear from Table 10–1 that the significant correlations are grouped in certain portions of the N matrix. The numerical abundances of calanoid and cyclopoid copepods through all stages show strong interrelationships indicating much coincidence in the peaks and troughs of abundance in the two species. The abundances of *Diaphanosoma* and *Bosmina* also overlap strongly with the abundances of copepods of all stages. *Conochiloides* and *Trichocerca* overlap strongly with cyclopoids but not with calanoids. The other four species of rotifer, *Moina,* and *Chaoborus* all show abundance patterns which bear no strong relationship with the abundance patterns of other species in general.

The cluster analysis for the N matrix (Fig. 10–2) breaks the community into four major groups. The first group contains two cladocerans (*Bosmina* and *Diaphanosoma*) plus about half of the copepod stages; the second group contains three rotifer genera (*Conochiloides, Trichocerca,* and *Tetramastix*) plus the remaining copepod stages; the third group contains the remaining four rotifer species, which are all relatively rare, and *Moina,* the rarest of the cladocerans; and *Chaoborus* clusters by itself as the fourth group. These groups are defined as might be expected on the basis of the distribution of significant correlations in the N matrix. The only point which might appear puzzling is the relationship between *Tetramastix* and the organisms of the second cluster. Although *Tetramastix* does not show any strong relationship in abundance with copepod stages, its abundance is highly correlated with that of *Conochiloides,* which brings it into the second cluster.

The ΔN Matrix

Of the 128 unique pairs of species or stages in the ΔN matrix, 30 (23%) produce statistically significant correlations. Approximately six of these could be accounted for by chance alone.

The distribution of statistically significant correlation values in the ΔN matrix is different from that of the N matrix. As with the N matrix, there is a tendency for significant correlations to occur between the copepod stages among both cyclopoids and calanoids, however. Another similarity with the N matrix is that the rotifer genus *Conochiloides* correlates in several cases with copepod stages, as does the cladoceran genus *Diaphanosoma.* In contrast to the N matrix, the overall relationship of *Bosmina* to copepods is much less pronounced. Also the changes in abundance of the genus *Chaoborus* show a strong positive relationship to the changes in abundance of cyclopoid copepods (but not calanoids) and to the rotifer genus *Polyarthra,* whereas in the N matrix there were no significant relationships for *Chaoborus.* As in the N matrix, the rarer rotifer species show little overall relationship to other groups of zooplankton, but in the ΔN matrix there is a more pronounced tendency for these species to show significant relationships to each other. Particularly notable is the significant negative correlation of ΔN values for the two *Keratella* species, which showed no significant relationship to each other in the N matrix.

The cluster analysis (Fig. 10–2) looks entirely different from the N matrix cluster analysis. At the bottom of the diagram, a group of five rotifer species including not only the dominant *Conochiloides* but also a number of rare species breaks away from the other zooplankton species. The other zooplankton species and stages are thoroughly mixed in the upper clusters of the diagram and do not show any evidence of pervasive taxonomic patterns such as those which appeared in the N matrix.

The *P* Matrix

In the *P* matrix, 30% of the correlations are significant (Table 10–2). As with the other two matrices, virtually all of the significant relationships are positive.

As might be expected from Figure 10–1, the overall pattern of significant relationships in the *P* matrix is intermediate between those observed in the *N* and Δ*N* matrices. There are strong interrelationships between copepod stages of both species, between copepods and the rotifers *Conochiloides* and *Trichocerca,* and to some extent between copepods and the two cladocerans *Diaphanosoma* and *Bosmina* and the genus *Chaoborus.* The rarer rotifers and *Moina* show only a few significant relationships.

The *P* cluster diagram can for purposes of discussion be divided into three groups of organisms. Toward the bottom, the rare rotifer species and *Moina* cluster together much as they did in the *N* matrix. Two cladocerans, one rare rotifer, and one copepod cluster together in the middle group, and the bulk of the copepod stages cluster together in the top group along with the predator *Chaoborus,* the dominant rotifer *Conochiloides,* and the rotifer *Trichocerca.*

The Δ*P* Matrix

In the Δ*P* matrix (Table 10–2), 28% of the correlations are significant. The proportion of significant negative correlations (5/36) is much greater than in any of the other three matrices, but still positive coefficients predominate.

As with all the other matrices, the copepods account for a substantial number of the significant correlation coefficients. Copepods show strong relationships to each other, to the rotifer genus *Conochiloides,* and to the dominant cladoceran, *Diaphanosoma.* As in the Δ*N* cluster, *Chaoborus* shows strong relationships specifically with cyclopoid copepods. A substantial number of relationships also appear in the matrix between calanoid copepods and the rarer species of rotifers.

The Δ*P* cluster analysis (Fig. 10–2) groups all of the copepods close together toward the uppermost portion of the diagram. This cluster suggests great similarities between the different stages of a copepod species and thus tends to segregate the species cleanly. *Chaoborus* and *Bosmina,* along with *Polyarthra,* are grouped with the copepods. Two of the bottom-most clusters account for all but two of the rotifer species, and an intermediate cluster accounts for two of the cladocerans and one relatively rare rotifer species.

Summary of Similarities

Table 10–3 summarizes the various kinds of ecological coherence that have been discussed above. The table is organized taxonomically to facilitate discussion. The number of groups into which the community can be divided is to some extent arbitrary but is generally based on the cluster diagrams (Fig. 10–2). Within any group in the table, copepods (if any) are always listed first, cladocerans are listed second, and rotifers are listed last. Rotifers are represented by a two-letter code derived from the first letter of the genus name and the first letter of the species name (Table 4–1).

It is reasonable to hypothesize that the various developmental stages of copepods should be dissimilar ecologically because of the great change in size and morphology of these species. There is a carryover of influence from one stage to another within a species, however, which cannot occur between species. In effect the carryover of biomass from one stage to another as development occurs forces a certain amount of ecological continuity between the stages despite adaptations of a physiological or morphological nature which tend to have the opposite effect. Table 10–3 shows that cyclopoid nauplii, copepodids, and adults show more similarity to each other in the timing of their demands on food resources and in their composite responses to growth and mortality factors than they do to any other single species or group of species. Thus in the context of community dynamics (ΔN, P, ΔP matrices), the developmental stages are tightly linked. Calanoid copepods show similar tendencies but are slightly more fragmented in their affinities than cyclopoids. Coherence between copepod stages does not extend to community structure, however (N matrix). The abundance peaks of the stages tend to be separated in time sufficiently to cause the stages to group with different species of zooplankton rather than with each other.

The affinity of the two copepod species is also of interest in view of their close taxonomic relationship and potential sharing of critical resources. In the three comparisons which have the greatest significance in terms of population dynamics (ΔN, P, ΔP matrices), the two copepod species have a definite tendency to segregate into different but adjacent groups, suggesting that they share some requirements but are still segregated from each other by some combination of adaptation and competition. When an overlap does occur in these groupings, it is between the cyclopoids and the latest developmental stages of calanoids. Since the largest calanoids do not overlap in size with any of the cyclopoids, this grouping pattern is suggestive of some kind of ecological exclusion of the small calanoid stages from the cluster occupied by cyclopoids. In terms of community structure, the species do not segregate at all clearly, which emphasizes the necessity of considering structure and dynamics on different bases.

The Cladocera are related in a complex manner to other species. In terms of community dynamics, however, the species have a strong tendency to

Table 10–3.   Summary of the Groupings Obtained in the Cluster Analysis Showing the Principal Basis for the Observed Similarities[a]

Basis of similarity	Evidence	Groupings			
		I	II	III	IV
Availability to predators, community structure	N matrix	Early cyclopoid and calanoid stages *Bosmina* —	Late calanoid stages *Diaphanosoma* —	Late cyclopoid stages — CD, TO, TS	— *Moina* KC, PV, KP, HI
Timing of demands on algal resources	P matrix	All cyclopoids — CD, TS	All calanoids *Diaphanosoma, Bosmina* PV	— *Moina* KC, KP, HI, TO	— — —
Overall ecological similarity in growth plus mortality	ΔN matrix	All cyclopoids, late calanoid stages *Diaphanosoma* PV	Early calanoid stages *Moina, Bosmina* KC	— — CD, KP, TS, TO, HI	— — —
Overall ecological similarity in growth	ΔP matrix	All cyclopoids, calanoid adults — —	Early and late calanoid stages *Bosmina* PV	— *Diaphanosoma, Moina* KC	— — CD, KP, TS, TO, HI

[a] Within the groups, the species are arranged taxonomically with copepods on top, cladocerans second, and rotifers third. Rotifers are represented by a two-letter code derived from the first letter of the genus name and the first letter of the species name.

group with each other or to occupy adjacent groups. The differences in the groupings when the emphasis is shifted from growth plus mortality (ΔN matrix) to growth more exclusively (ΔP matrix) are particularly interesting. The predator *Chaoborus* shows a very strong preference for cladocerans in Lake Lanao (Lewis, 1977a), which would provide a foundation for the hypothesis that these species are particularly affected by the timing of predation mortality. Table 10–3 shows that the inclusion of mortality as a basis for grouping radically shifts the grouping affinities of the cladoceran species. It would thus appear that in the absence of strong selective predation on Cladocera, the cladocerans would have a strong affinity with rotifers and late developmental stages of calanoids.

The first entry of Table 10–3 shows that the cladocerans are temporally well separated in terms of their availability to predators. This may well be adaptive in the sense that it reduces the aggregate number of cladocerans at any given time. *Chaoborus* in Lake Lanao reduces its selectivity for Cladocera when the density of Cladocera is low (Lewis, 1977a).

The rotifers are a remarkably coherent group both in community dynamics and community structure. *Polyarthra vulgaris* is most unusual in terms of dynamics, as it shows a tendency toward affinities with calanoids and cladocerans rather than with other rotifers. In comparing the similarities based first on growth plus mortality and then more exclusively on growth, it is apparent that there is little difference with regard to rotifers in the grouping pattern, which suggests that growth control mechanisms are of overriding importance in determining rotifer dynamics. It is interesting that the dominant rotifer, *Conochiloides*, shows no similarity whatever to cyclopoid copepods in terms of growth control mechanisms and yet its timing in the exploitation of algal resources is very similar to that of cyclopoid copepods (P matrix). This is most logically explained by the vast difference in response and turnover times for the two species. In terms of community structure, association of rotifer species with other species is minimal.

The groupings that have been discussed here allow only minimal identification of mechanisms that might be separating the groups. These mechanisms will be the principal concern of the subsequent analysis.

Chapter 11

Zooplankton Mortality Rates

Approximations were made of the mortality rate for each species and developmental stage in the Lake Lanao zooplankton community. The method of computation differs somewhat between species as indicated below.

Computations: Copepods

Copepod mortality was determined from a computer simulation incorporating the egg ratios and development rates that have already been documented. The simulation is as follows.

1. The sampling period (7 days) is divided into 70 equal time intervals of 0.1 day each, and the life history of the copepod is divided into age classes of 0.1 day (approximately 500 age classes). The mortality of the population is determined separately for each of the 0.1-day age classes and each of the 0.1-day time intervals. Determination of mortality over these short time intervals approximates without significant error (less than 5%) the instantaneous mortality.

2. The population abundance and age structure at the beginning of any 0.1-day interval are determined from the sampling data at the beginning and the end of the 7-day sampling period within which the 0.1-day interval falls. Age structure and abundance for a particular 0.1-day interval are computed as the time-weighted average of known values for the beginning and end of the appropriate 7-day sampling period.

3. The effect of development over any 0.1-day interval is determined for the entire population simply by advancing the age of all organisms in the population by 0.1 day (including developing eggs, a proportion of which will hatch). This represents the theoretical abundance and age structure without mortality. The difference between this simulated population and the actual population at the end of the 0.1-day interval is due to mortality over the interval. The actual abundances and age structure are computed as time-weighted averages of samples at the beginning and end of the 7-day period within which the 0.1-day interval lies. Mortality from a total of 70 0.1-day computations is accumulated over the period of 7 days spanning two samples to give the total mortality in each age class for the week.

4. The number of organisms entering the population as a result of reproduction over any of the 0.1-day intervals is determined from the birth rate, which is computed from the ratio of eggs to females in the population and the development rate of eggs. The birth rate is not assumed to be constant over the 7-day period between samples, but is assumed to vary in a linear fashion between the observed birth rate at the beginning of the period and the observed birth rate at the end of the period.

5. The mortality rates for each 7-day period are expressed as numbers of organisms per liter per day and also as percentage mortality per day.

Since the procedure is based on estimates of development rate and abundance, errors may result from any inaccuracy in either of these variables. The procedure can even yield negative mortality rates. This is particularly likely when the true mortality rate is very low, as any slight underestimation of the development rate or overestimation of the population size at the end of the time interval might easily exceed the actual mortality and thus indicate a negative mortality. Thus while negative mortality rates are impossible in nature, they are an expected by-product of the simulation procedure on a statistical basis and should not be arbitrarily eliminated. The analysis will focus on trends and averages rather than the specific mortality rate for any particular week, hence negative mortality rates are retained wherever they are produced by the estimation procedure. Results of the computations are summarized in Table 11–1 both in terms of absolute mortality (individual per liter per day) and relative mortality (percentage per day).

Computations: *Chaoborus*

Chaoborus mortality was computed by a procedure identical to the one used for copepods. Approximation of the number of organisms entering the population by means of reproduction is more difficult for *Chaoborus*, however,

Table 11-1. Mean Absolute Mortality Rates and Weighted Mean Relative Mortality Rates of Lake Lanao Zooplankton over a 65-Week Period, and Their Modified Coefficients of Variation (CV′)[a]

Species/stage	Mortality (individual per liter per day)	Coefficient of variation (%)	Mortality (% per day)	Coefficient of variation (%)
Copepods				
Thermocyclops				
Nauplii	1.43	34	1.6	43
Copepodids	5.18	98	15.6	53
Adults	0.57	45	5.0	48
Tropodiaptomus				
Nauplii	0.66	64	18.5	52
Copepodids	0.21	29	9.6	39
Adults	0.18	53	8.0	51
Rotifers				
Conochiloides	4.61	111	26	16
Hexarthra	0.66	203	15*	—
Polyarthra	0.14	263	17*	—
Keratella procurva	0.18	170	26*	—
K. cochlearis	0.71	152	18*	—
Trichocerca	0.04	263	15*	—
Tetramastix	0.54	123	12	136
Cladocera				
Diaphanosoma	0.97	93	20	26
Moina	0.11	184	26*	—
Bosmina	0.24	267	16*	—
Diptera				
Chaoborus				
Instar I	0.00200	84	2.2	118
Instar II	0.00140	46	5.4	35
Instar III	0.00126	40	4.5	57
Instar IV	0.00073	31	5.6	69

[a] Asterisks, subject to special interpretation (see text).

because adults and eggs were not counted. Pupa and egg development were assumed to require 1 day each (Juday, 1921; Berg, 1937; Parma, 1971). Although much longer development times may occur in cold water (Parma, 1971), the warmth of Lake Lanao promotes very rapid development. Pupae were present only in very low numbers, suggesting that the estimate of 1.0 day for pupal development is quite reasonable. Each female *Chaoborus* was assumed to lay 400 eggs. Parma (1971), who has made the most careful study of egg numbers, found that the average number of eggs laid by *Chaoborus flavicans* in a Dutch pond was 449.

Adults can live as long as 6 days (Parma, 1971) but probably complete reproduction quickly in nature. For simulation purposes, adults were assumed to lay eggs within 24 h of emergence. The number of adult female organisms which had not yet laid eggs at the beginning of any particular time interval was approximated from the observed number of larvae in the last instar of development and from the known development rate and mortality rate of that instar.

The estimates dealing with adults, eggs, and pupae are obviously very approximate by comparison with the estimates dealing with larvae. The issue is further complicated by the tendency of *Chaoborus* to move over long distances prior to depositing eggs. Berg (1937) mentions that adult *Chaoborus* have a tendency to lay eggs around the shoreline, but Parma (1971) notes that this is not always true. Adult *Chaoborus* in Lake Lanao swarm heavily around the shoreline, and Lewis (1975) has documented a definite trend in age distribution of organisms which suggests that more eggs hatch in the near-shore areas than in the open waters of the lake. As a result, the estimates of mortality in adults and of individuals entering the population are only rough approximations, but the estimates of larval mortality are much more reliable.

Computations: Rotifers and Cladocera

Mortality estimates for rotifer and cladoceran populations were obtained by a method similar to that used for copepods and *Chaoborus*. The simulation was identical except that (1) only two developmental stages (egg and adult) were recognized rather than the more numerous stages of copepods and *Chaoborus;* and (2) egg ratios were set to the annual average when the population reached such low levels that counting error produced an expected coefficient of variation greater than 10%.

Some special limitations must be placed on the interpretation of relative mortality rates in these species characterized by only two distinguishable developmental stages. As indicated in the discussion of calculation methods, the egg ratio on a given date for a given rotifer or cladoceran species is set to the annual mean egg ratio if on that date insufficient numbers of organisms were present to give better than 10% expected coefficient of variation due to counting error. This restriction was not applied to copepods because their life history is much longer and is divided into more stages so that inaccuracies in the egg ratios do not bias the estimate much, and because large numbers of copepods were always available for counting. Furthermore, the restriction had virtually no effect on the estimates of mortality for *Conochiloides, Tetramastix,* or *Diaphanosoma* (the three most abundant species of rotifer and Cladocera), which were almost always present in sufficient abundance to meet the 10% criterion. The remaining five rotifer species and two cladoceran species, however, were rare enough on many occasions to require use of the mean egg ratio rather than the egg ratio at the time of sampling.

The reason for the restriction, of course, is to minimize distortions in the apparent egg ratio on dates when too few eggs and females are available for accurate counting. Absolute mortality rate is only partially affected by use of an average egg ratio, as the absolute mortality rate is dependent on both the abundance at the time the mortality rate is estimated and upon the egg ratio from which the birth rate is estimated. The relative mortality rate, however, is greatly influenced by the use of a mean egg ratio insofar as estimated relative mortality is entirely dependent upon the estimated birth rate, which is in turn dependent upon the egg ratio. For a species which is rare most of the year, the egg ratios are often set to a fixed value equal to the mean value for the year. This means that the variation in relative mortality rates of rare species cannot be studied with any great degree of confidence. Consequently the coefficients of variation for relative mortality in these species are not given in Table 11–1. Also the mean relative mortalities are marked with an asterisk to indicate they are estimated from a smaller data base than for other species. Although the restrictions are less binding on the absolute mortality rates, the relative mortality rates are much more likely to reveal mortality mechanisms and must be used in mortality analysis.

Measurements of absolute and relative mortality in Table 11–1 are in most cases accompanied by a relative measure of variation derived from the coefficient of variation. The coefficient of variation is not a good measure of relative variation when a variable fluctuates about 0, since in this case it will be unduly influenced by any minor changes in the mean. Theoretically this is not a problem with mortality rates, since mortality rates must always be positive. In actuality, mortality rates which are close to 0 may sometimes be estimated as a negative value due to the presence of error variance in the estimates. For this reason, the coefficients of variation were computed as follows:

$$CV' = [s/(\bar{x} + k)] \cdot 100$$

where k is exactly large enough to make 95% of the values of x positive. This correction makes the coefficients of variation more truly comparable between species.

Copepod Mortality Rates

For cyclopoid copepods, the mortality rate changes substantially from one developmental stage to another. Figure 11–1 shows the weighted average percentage mortality rate throughout the life history. Points on the graph

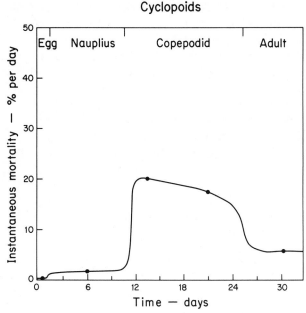

Figure 11–1.   The mortality of *Thermocyclops hyalinus* in Lake Lanao throughout the developmental cycle (weighted average over a 65-week period).

represent the mean mortalities for eggs, nauplii, copepodid stages I–III, copepodid stages IV–V, and adults. Although the nominal resolution of the data was actually much finer than this, computations of mean mortality for very brief periods of the life history are unduly influenced by sampling errors and errors in the estimation of development time, so the interpretation is based on mean mortality over more extensive periods of the life history during which the organisms vary little in morphology.

Mortality rate for cyclopoid eggs is nearly zero. *Chaoborus* does eat adult cyclopoids, but the eggs are so rapidly synthesized (because of small size) and develop so quickly that relative impact of predation on the eggs themselves is low. Lewis (1977a) has shown that ovigerous females are no more vulnerable than males or copepodid stages of cyclopoids and may even be less vulnerable than some of these stages. Copepod eggs are ingested along with female copepods but cannot be successfully macerated by *Chaoborus*. Eggs may thus remain viable after regurgitation from a *Chaoborus* crop, although this has not been tested experimentally. Figure 11–1 also implies that sources of mortality such as disease and parasite invasion do not affect significant numbers of eggs in the field populations.

The mortality of naupliar stages is extremely low, but the mortality of copepodids is quite high. Mortality among copepodids tends to decline with size and reaches a minimum in adult copepods. As will be demonstrated more fully in connection with the analysis of predation, the shape of the curve in Figure 11–1 is easily explained on the basis of differential vulnerability of developmental stages to predation. The nauplii practice a cryptic swimming behavior which renders them all but immune to predation by *Chaoborus* (Lewis 1977a, Gerritsen 1978). The swimming behavior of the copepodid renders it highly vulnerable to predation, but as the copepodid increases in size an increasingly smaller percentage of the predator population is able to capture, hold, and ingest it. This accounts for the decline in mortality of copepodid stages with age, and particularly for the low mortality of adult males and ovigerous females, which are the largest individuals in the population.

Figure 11–2 shows the change in mortality during the life history of calanoid copepods. The pattern is completely different from that of cyclopoids. Mortality is low in the egg stage by comparison with other life history stages, probably for reasons similar to those already given for cyclopoids. In contrast to cyclopoids, however, the nauplii suffer very high mortality. The nauplii thus either do not possess or do not profit from a cryptic swimming behavior similar to that of cyclopoids. The reduced losses of copepodid stages may be accounted for by their large size, which renders them less available to small *Chaoborus,* or to behavioral escape mechanisms.

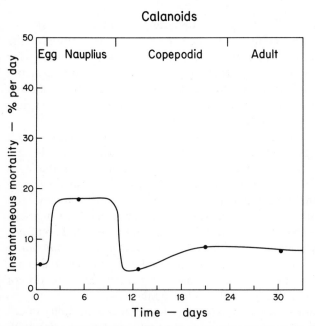

Figure 11–2. The mortality of *Tropodiaptomus gigantoviger* in Lake Lanao throughout the developmental cycle (weighted average over a 65-week period).

The difference in percentage mortality through comparable stages of cyclopoid and calanoid copepod populations in Lake Lanao is evident not only in the mortality calculations but also in the population structures themselves. The cyclopoids show a very high ratio of nauplii to copepodids (Table 4–1), whereas the calanoids show a much lower ratio of nauplii to copepodids despite their very similar life cycle and development schedule.

Chaoborus **Mortality Rates**

Chaoborus mortality is summarized in Table 11–1 and Figure 11–3. Instantaneous mortality rates are generally not so high as for copepods or other zooplankton, but the duration of development is considerably longer. Mortality rate in instars II–IV is very constant, but mortality rate in instar I is only about half of that observed in later instars. The increase in mortality rate of larger instars could well be due to selective feeding by vertebrate predators on larger organisms, as fish predation is the major source of larval mortality. The adult and egg stages experience much higher relative mortality than the largest larval stages, as would be expected. The adult in effect sacrifices itself in order to lay eggs.

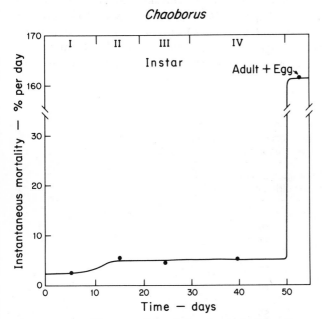

Figure 11–3. The mortality of *Chaoborus* in Lake Lanao throughout the developmental cycle (weighted average over a 65-week period).

Rotifer and Cladoceran Mortality Rates

The absolute mortality rates for rotifer populations vary in relation to the abundance of the species, as expected (Table 11–1). The relative mortality rates are not nearly so variable as one might expect. Despite the small size of the rotifers, their weighted average relative mortality rates do not greatly exceed those of the more vulnerable developmental stages of copepods. Few annual average relative mortalities from field populations are available for comparison. Rey and Capblancq (1975) showed average loss of 14% per day for *Polyarthra vulgaris,* which is very close to the average for rotifers in Table 11–1.

Cladoceran mortality rates are on a relative basis very similar to rotifer mortality rates (Table 11–1). The two larger cladoceran species sustain somewhat greater relative mortality than the small *Bosmina*.

Overview of Mortality Rates

Higher coefficients of variation for absolute mortality than for relative mortality in a given zooplankton category would be evidence of substantial density dependence in mortality. This is the case for some categories and not for others. Mortality in copepods, for example, does not appear to be markedly density dependent except perhaps for *Thermocyclops* copepodids, but mortality for the rotifer *Conochiloides* appears to be highly density dependent.

Among copepods, those stages which are characterized by the highest relative mortality rates show the highest coefficients of variation for both abso-

Table 11–2. Average Percentage of Organisms Surviving from Birth to the End of Given Developmental Stages in the Three Zooplankton Species with Extended Life Histories

Species/stage	Percentage surviving to end of developmental stage
Cyclopoids	
Nauplius 6	86
Copepodid V	11
Calanoids	
Nauplius 6	27
Copepodid V	10
Chaoborus	
Instar II	50
Instar IV	17

lute and relative mortality. This is probably because the mortality of these stages is principally caused by predation, which is unevenly distributed through time.

Relative variation in the mortality of the predator *Chaoborus* is lower than variation in most of the herbivore categories when mortality is expressed in absolute terms. When mortality is expressed in relative terms, the relative variability of *Chaoborus* mortality is in the same range as the relative variability in herbivore mortality.

Table 11–2 shows the combined effect of relative mortality and duration of development on survival to age of first reproduction for the three species with extended periods of development (cyclopoids, calanoids, and *Chaoborus*). The percentage of surviving copepods at the time of the last molt is remarkably similar in the two copepod species (ca. 10%). These figures are also very similar to the ones obtained by Bosselmann (1975) for *Eudiaptomus* in Lake Esrom (ca. 10%). Broader investigation of mortality rates might well show general similarities among copepod populations in this respect.

Chapter 12

Zooplankton Mortality Mechanisms

Mortality among zooplankton can be broken down into two large categories: (1) predation, and (2) all other causes, including disease, physical stress, parasitism, starvation, and senescence. Quantification of the relative importance of these two categories of mechanisms for the various stages and species included in the Lake Lanao zooplankton is possible because the rates of predation can be computed.

Predation Mortality

All predation on zooplankton in Lake Lanao must be attributed either to fish or to *Chaoborus,* since all of the other zooplankton species are herbivorous. Lake Lanao is populated by a species swarm of endemic cyprinids plus some introduced fish species, including predominantly the piscivorous white goby, *Glossogobius giurus.* The lake lacks a significant population of fish adapted to feed on herbivorous zooplankton in the pelagic zone. Juvenile fishes and some of the smaller cyprinid species which are zooplanktophagic are limited in distribution primarily to the littoral areas, where they probably have a significant impact on zooplankton populations. The abundance of juvenile fish and the smaller cyprinid species in open water is so low, however, that this source of predation is not significant for herbivorous zooplankton in mid-lake. Some of the larger cyprinid species, which are not subject to intensive predation by the piscivorous goby, are found in open water but do not appear to be effective zooplankton feeders except on *Chaoborus,* which appears in large numbers in the stomachs of some species.

From these generalizations it is possible to make some simplifying assumptions concerning the impact of fish on the zooplankton community. *Chaoborus* larval mortality is almost certainly accounted for entirely by fish predation. *Chaoborus* shows no evidence of significant cannibalism in Lake Lanao (Lewis, 1977a), although cannibalism does occur in at least some temperate populations (Parma, 1971). Given the small amount of fish predation on zooplankton other than *Chaoborus,* mortality of herbivorous zooplankton must be accounted for entirely by *Chaoborus* predation or by factors other than predation. The feeding habits of *Chaoborus* have been studied in detail, so it is possible to compute predation and thus separate predation losses from other losses for the herbivorous zooplankton.

Computation of predation mortality for herbivores requires knowledge of the feeding rate of all *Chaoborus* instars on all herbivore categories. There is a diel rhythm in electivity values for Lanao *Chaoborus* (Lewis, 1977a), hence the calculations must be made separately for day and night. The electivity values for diurnal and nocturnal feeding across all *Chaoborus* instars, when combined with the abundance matrix for herbivores, can be used to compute the percentage by number or biomass of each herbivore category in the *Chaoborus* diet for each of the 65 weeks.

Given the percentage composition of the *Chaoborus* diet for each week, total predation mortality of each herbivore category can be computed if the feeding rate is known throughout development for *Chaoborus.* The feeding rate for *Chaoborus* can be computed from (1) the growth rate, which is known (Fig. 8–6), and (2) the growth efficiency. Swift (1976) reports an assimilation rate of 65% for *Chaoborus,* which will be used here. Relationship of growth to assimilation has not been studied in any *Chaoborus* species, but it seems reasonable to adopt the percentage determined by Cummins et al. (1969) for the zooplankton predator *Leptodora* (30%). These two percentages are applied to the age-specific growth curve to obtain the age-specific feeding rates expressed as a percentage of *Chaoborus* body weight. The size –abundance–time matrix for *Chaoborus* incorporating all 65 weeks of the study period and all 52 separate *Chaoborus* size categories is then applied to the category–abundance–time matrix for herbivores to yield the predation losses of all herbivores. Results are expressed in absolute terms (individuals/liter/day) and in relative terms (percentage/day).

The computations are subject to all of the errors that have already been mentioned for estimates of abundance and growth. In addition, two other sources of error should be mentioned. (1) Electivity values for *Chaoborus* in the Lake Lanao community vary according to the relative abundance of available foods, even for a specific stage in the *Chaoborus* life cycle at a specific time of year. Lewis (1977a) has shown that *Chaoborus* lowers its electivity for preferred foods when such foods reach very low density, presumably because it becomes counterproductive to seek or wait for the food under these circumstances. It is not known, however, whether this relaxation of preference occurs suddenly or slowly. The use of average electivities could thus cause some inaccuracy. (2) Additional potential inaccuracies in the calculations arise from the use of a single estimate of assimilation efficiency and of the assimilation–growth relationship. The true percentages are likely to vary some according to the type of food being consumed and the age of the predator.

The results of the computations are listed in Table 12–1. The computations were made for the 16 herbivore categories listed in the generalized mortality summary (Table 11–1) except that immature copepodids and adult copepods were combined because they could not be distinguished in the diet of *Chaoborus*. Fourteen herbivore types were thus considered. Of these fourteen, five, all of which are rotifer species, are never found in the diet of *Chaoborus* (Lewis, 1977a). Obviously the mortality due to *Chaoborus* predation in these five categories is 0, hence all of their mortality must be accounted for by other means. These categories are omitted from Table 12–1, which shows only the nine herbivore categories that are subjected to detectable predation by *Chaoborus* even though the total predation loss may be very minor (cyclopoid nauplii).

Table 12–1 shows that the total daily mortality of herbivores which are known to serve as *Chaoborus* food (32 μg wet weight/liter/day) very closely matches the total computed *Chaoborus* food requirements (34μg wet weight/liter/day). Since these two estimates are entirely independent of

Table 12–1. Mean Mortality of Herbivores Attributable to *Chaoborus* Compared with Mean Total Mortality

Species/stage[a]	Predation mortality (μg/liter/day) (wet)	Total mortality (μg/liter/day) (wet)
Cyclopoid nauplii	0.05	0.43
Cyclopoid copepodid/adult	12.83	18.39
Calanoid nauplii	0.44	1.12
Calanoid copepodid/adult	13.75	6.44
Keratella cochlearis	0.15	0.05
Keratella procurva	0.01	0.02
Diaphanosoma	5.20	3.88
Moina	0.35	0.66
Bosmina	1.34	0.58
Total	34.13	31.57

[a] Species and stages which experience negligible mortality due to *Chaoborus* predation are excluded from the table.

each other, the close agreement confirms that (1) sources of predation other than *Chaoborus* are not important for herbivorous zooplankton species, and (2) predation is by far the most important source of mortality for the nine categories which are listed in Table 12–1 and probably accounts for virtually all loss of biomass in these categories. The close agreement of estimates furthermore supports the validity of the estimates of *Chaoborus* feeding rates and growth rates as well as the overall mortality computations for herbivore populations.

For individual herbivore categories in Table 12–1, the agreement between predation mortality and total mortality is acceptable but not nearly so close as for the total. The major source of inaccuracy is undoubtedly the estimate of electivities for *Chaoborus* of various sizes, as the mean electivity values are unquestionably subject to numerous sources of variation which could not be well quantified. Furthermore, the analysis of *Chaoborus* foods, which was based on studies of crop contents, could not be nearly so precise as the analysis of total mortality rates because the exact size of organisms within *Chaoborus* crops could not be determined. These sources of error tend to balance in the totals, but are a source of variation in the estimates of predation loss rate for individual herbivore categories.

The herbivores can be divided into two major groups according to the importance of predation in explaining mortality. The first group is composed of all the species listed in Table 12–1. Predation seems to account satisfactorily for all mortality in these species. Within this group, the exact mortality rate caused by predation varies enormously according to the variation in electivity of *Chaoborus* for various food items. This is in turn reflected in the variation of mean relative mortality rates in Table 11–1. The second major group of zooplankton herbivores consists of the five rotifer species which do not appear at all in the diet of *Chaoborus* and which nevertheless sustain relatively high mortality rates.

The existence of five herbivorous zooplankton species in which mortality must be accounted for entirely by mechanisms other than predation is remarkable and deserves further consideration. For zooplankton species which have a very low average numerical abundance, it is possible that an amount of mortality which was significant to the species in question did not appear in the dietary analysis of *Chaoborus* simply because the percentage of such items in the food is very low. An example would be the rotifer *Trichocerca*. For at least two of the species, however, the evidence that substantial mortality must occur by mechanisms other than predation is very definite. The abundances of *Conochiloides* and *Tetramastix* are so high (Table 4–1) that these items could not be overlooked if they were truly included in the diet of *Chaoborus* in significant numbers.

Mechanisms of Mortality Other Than Predation

Although five rotifer species suffered significant mortality that cannot be accounted for by predation, no direct observational evidence is available to suggest mechanisms for this unexplained mortality. Mechanisms might include disease, parasitism, inadequate food supply, physicochemical stress, or senescence, all of which are difficult to document. There was no evidence of disease or parasitism, so the role of these factors will be tentatively considered minimal. It is possible to test for relationships between mortality and (1) the quality and quantity of the food resource, (2) physicochemical stress, and (3) hidden predation by large herbivores which might feed occasionally on small herbivores.

The basis for analysis of mortality mechanisms is the mortality–time matrix, where mortality is expressed in relative terms (percentage/day). Of the five rotifer species which should be analyzed for unexplained mortality, only two (*Conochiloides, Tetramastix*) are sufficiently numerous to support detailed analysis.

1. Mortality versus Food Supply

Weekly abundance estimates are available for each of the 70 phytoplankton species found in the limnetic zone, as described elsewhere (Lewis, 1978b). The data can be expressed either in terms of plankton units per ml (plankton units are independent biomass units, or individuals), or in terms of biomass per ml of water (as computed from cell volume). These abundance measurements for individual phytoplankton species and aggregates of species were compared with the mortality rates for *Conochiloides* and *Tetramastix*.

In all of the statistical comparisons, the abundance of phytoplankton species or species aggregates for a given week was obtained by summing the abundance data for the beginning and the end of the week and dividing by two. The general hypothesis guiding the analysis in all cases is that unexplained herbivore mortality is statistically related to the availability of critical food resources within the phytoplankton. The analysis was conducted in several separate stages as described below.

A. Relationship between Zooplankton Mortality and Total Phytoplankton Abundance

Pearson Product–Moment Correlation Coefficients were obtained between total phytoplankton abundance and the relative mortality rates of *Conochiloides* and *Tetramastix*. Initially none of the variables was transformed. A causal relationship between herbivore mortality and a nutritionally critical minimum of total phytoplankton would be indicated by a negative relationship between herbivore mortality and total phytoplankton abundance (although this relationship need not be linear, of course). No such relationship was found. The test was repeated after log transformation of the variables and again showed no significant negative correlations. The test was also repeated with incorporation of time lags of 1 and 2 weeks between phytoplankton abundances and zooplankton mortality and the result was the same. The test thus provided no evidence for connection between total phytoplankton abundance and the unexplained mortality in herbivores. This is not surprising, since the total phytoplankton abundance in Lake Lanao is extremely high throughout the year (Fig. 7–1), which would suggest that any deficiencies in the food supply must derive from the fact that part or all of the algal biomass is undigestible or non-nutritious at certain times of the year.

B. Relationship between Herbivore Mortality and the Abundance of Various Taxonomic Groups of Phytoplankton

The phytoplankton can be grouped at the class or division level into the following major categories: Cyanophyta, Chlorophyta, Bacillariophyceae, Dinophyceae, and Cryptophyceae. The phytoplankton actually contains two other groups (Euglenophyta and Chrysophyceae), but both of these are comprised of only a few very rare species.

The abundance of each of the five major taxonomic groups, expressed in terms of plankton units per ml, was tested for correlation with mortality in *Conochiloides* and *Tetramastix*. The tests were done with and without log

Table 12–2. Relationships between Indicators of Food Quality or Quantity and Mortality Rates of the Two Abundant Rotifer Species Showing Significant Mortality Unexplained by Predation

	Relative mortality—percentage/day	
Resource supply indicator	*Conochiloides*	*Tetramastix*
1. Total phytoplankton abundance (μm^3/ml or plank units/ml)	None	None
2. Abundance of major algal taxa (μm^3/ml or plank units/ml)	Greens, dinoflagellates, cryptomonads (+)[a]	Dinoflagellates (+)
3. Abundance of specific algal size fractions		
a. Volume per biomass unit (μm^3/ml)	60–120, 320–800 (+)	None
b. GALD (μm)	1.0–6.5, 12.5–22.0 (+)	None
c. SGALD (μm)	2.2–4.5, 11.5–27.0 (+)	None
4. Abundance of individual algal species	Greens 13, cryptomonads 2, other divisions 3 (+)	None

[a] Plus indicates positive correlation.

transformation of variables and with and without 1-week time lags. The results were qualitatively similar in all cases, but log transformation appears most justifiable in view of the frequency distributions, so interpretation is based on the transformed data.

There are no significant negative correlations (Table 12–2), but relative mortality of both species is positively related to dinoflagellate abundance and *Conochiloides* mortality is positively related to abundance of greens and cryptomonads as well.

C. Relationship between Zooplankton Mortality and Abundance of Phytoplankton Groups Based on Size

Since phytoplankton size is one factor which might affect its availability to a given herbivore, size was used as a criterion for aggregating phytoplankton species. The entire list of 70 species was split into 6 groups containing 11 or 12 species each. The groups were based on the size of an average biomass unit (μm^3/ml) for each phytoplankton species. Data on the size of average biomass units for all species were available from another study (Lewis, 1978a). The species span a volume range of 2 to 50,000 μm^3 per plankton unit, but 90% of the species had volumes between 3 and 2500 μm^3 per plankton unit. The volume of an average plankton unit for the entire 70 species was about 75 μm^3.

A correlation analysis of the six different size groups of phytoplankton with the mortality rates of the two herbivore categories showed no significant negative correlations and two significant positive ones, both involving *Conochiloides* (Table 12–2). Repetition of the tests after log transformation and with the incorporation of 1-week lags did not result in any meaningful increase in the number of significant correlations.

The entire analysis was repeated on the basis of linear dimensions rather than volume as a criterion for categorizing the phytoplankton. Once again the 70 species were split into six groups, but this time on the basis of greatest axial linear dimension (GALD, Lewis, 1976) of average biomass units for each phytoplankton species. The results were very similar to those obtained on the basis of volume per plankton unit (Table 12–2).

A third and final kind of size grouping of the phytoplankton was based on the second greatest axial linear dimension (SGALD) for each of the 70 species. The rationale for this separation was that the ingestibility of elongate biomass units might be more dependent on their greatest width than their length. The results, however, were essentially the same as in the grouping based on GALD.

The analysis of phytoplankton size fractions using any of the three criteria is on the whole not very revealing. In all three cases, two significant positive correlations appear for *Conochiloides* but these are with noncontiguous size groups of intermediate dimensions. If size is truly an important feature in itself, one would expect significant correlations to appear with contiguous size groups, particularly those groups of extreme size at one end or another of the spectrum. Furthermore, there are no significant correlations for *Tetramastix* at all. The results thus suggest that the significant correlations which were observed are the result of the operation of some factor other than size.

D. The Relationship between Zooplankton Mortality and Abundance of Individual Phytoplankton Species

Tests were also made for significant relationships between mortality of *Conochiloides* and *Tetramastix* and the abundance of the 70 individual phytoplankton species. As with the previous tests, the results were obtained with or without log transformation and time lags of one week. The analysis based on log transformation of all variables and no time lag seemed best justified from frequency distributions and general strength of correlations, hence the interpretation is made on this basis.

One additional problem occurs in connection with the analysis of individual species insofar as a large number of individual correlation tests will produce a small number of significant results even if no relationship exists in the data. These spuriously significant results can increase in number under some conditions if there is a large amount of autocorrelation in the data, as previously indicated in connection with the analysis of species similarities. The phytoplankton abundance data show a relatively low amount of autocorrelation, however (Lewis, 1978d). Without any major autocorrelation effects, a series of 70 correlations of phytoplankton abundance with the mortality rates for one of the zooplankton species would be expected to produce $0.05 \times 70 = 3.5$ significant results by chance alone, half of which would be positive and half negative, if there were no relationship whatever between the variables. We must therefore assume that there is a background level of approximately two positive and two negative significant correlations for each of the two zooplankton species. Interpretations must obviously not be made unless the number of significant correlations is substantially above this background level.

As with most of the previous tests, *Tetramastix* showed no evidence of significant positive or negative correlations with phytoplankton abundance except as would be expected by chance alone. Table 12–2 thus indicates no relationships for this species.

For *Conochiloides*, the most abundant rotifer species, the number of negative relationships is not above the background level, but the number of positive relationships is substantially above the background level (18). As indicated in Table 12–2, most of these positive relationships involve the Chlorophyta. It is also noteworthy that although there are only two cryptophyte species, both show a positive relationship to relative mortality of *Conochiloides*. The remaining few relationships are scattered and are difficult to separate from background. The pattern of positive correlations thus reinforces the result obtained in the analysis of phytoplankton taxonomic groups. Abundance of cryptophytes and green algae coincides with unfavorable conditions for *Conochiloides,* as indicated by high relative mortality rates.

Table 12–2 summarizes all of the attempts to relate unexplained mortality in *Conochiloides* and *Tetramastix* to quantity and quality of food supply. For *Conochiloides,* there does appear to be a significant relationship between certain categories of algae (greens, cryptomonads, and possibly dinoflagellates) and high relative mortality rates. A similar relationship between *Tetramastix* and the dinoflagellates could not be confirmed by the analysis of individual phytoplankton species. Although no absolute proof is possible on the basis of this information, the data do suggest strongly that food resources are qualitatively inadequate for these rotifers at certain times in the successional cycle when the phytoplankton is dominated by groups of species which are either difficult to capture or have poor food value for these rotifers. It will be possible to reinforce these conclusions on the basis of additional information from the section on growth control mechanisms.

2. Mortality versus Physical–Chemical Stress

Physical–chemical stress is a second possible source of unexplained mortality. Good documentation is available on the physical and chemical variables affecting the plankton community over the period of study (Lewis 1973a, 1974, 1978b), so it should be possible to detect any important relationship between unexplained mortality and physical or chemical stresses. It is difficult to advance an interesting null hypothesis, however, because the degree of variation in most of the physical and chemical variables is so slight that it would be extremely unlikely for any such variations to result directly in mortality of zooplankton organisms.

Only one possibility seems worthy of testing. The considerable seasonal variation in amount of turbulence in the water column might affect the rotifers in particular insofar as it could prevent these species from maintaining an optimal position in the water column or could demand large amounts of energy for the maintenance of position. This hypothesis was tested by comparison of the relative mortality rates for *Conochiloides* and *Tetramastix* with the depth of mixing over successive weeks in the study period. Linear, log-transformed and time-lagged correlations were computed. These gave qualitatively similar results. The log-transformed correlations seem to be best justified on the basis of the frequency distributions. Table 12–3 summarizes the results.

Table 12–3 shows that depth of mixing is significantly related to mortality rates in *Conochiloides.* The sign of the relationship between these two variables is the opposite of the sign expected on the basis of the initial hypothesis, however. The negative relationship between mortality and mixing depth for *Conochiloides* suggests that these species experience

Table 12–3.   Pearson Product–Moment Correlation Coefficients between Measures of Potential Hidden Sources of Mortality and Relative Mortality (Percentage/Day) for Herbivore Categories Showing Significant Amounts of Unexplained Mortality

	Relative mortality—percentage/day	
Potential source of mortality	*Conochiloides*	*Tetramastix*
Depth of mixing	−0.26[a]	NS[b]
Abundance of other herbivores		
Cyclopoid nauplii	NS	NS
Cyclopoid copepodid/adult	NS	NS
Calanoid nauplii	NS	NS
Calanoid copepodid/adult	NS	NS

[a] $P < 0.05$.
[b] Not significant.

minimal mortalities when mixing is pronounced. The original idea is therefore rejected and the results do not suggest any simple alternative hypothesis. A significant negative relationship may be due to correlation of mixing with controlling variables that have not yet been identified or may signify some unexpected mechanism by which minimal mixing directly increases mortality. Even if such a mechanism does exist, its importance in accounting for variations in relative mortality is not very great since the intensity of the relationship is unimpressive.

Since the only reasonable hypothesis involving physical or chemical stress does not appear to explain any significant amount of the unexplained mortality in the rotifers, the influence of physical and chemical factors as a whole on unexplained mortality must be considered minimal on the basis of the available evidence.

3. Hidden Predation from Large Herbivores

One additional possible reason for unexplained mortality in the rotifers derives from the possibility that the larger herbivores, particularly cyclopoid copepodid/adults and calanoid copepodid/adults, might ingest rotifers in sufficient amounts to affect population dynamics. Although the examination of gut contents of the two copepod species from Lake Lanao shows unequivocally that they are herbivorous, and the mouthpart morphologies confirm this conclusion, it is not impossible that such herbivorous species ingest small heterotrophs in significant amounts. The relationship between abundance of cyclopoid nauplii, cyclopoid copepodid/adults, calanoid nauplii, or calanoid copepodid/adults with the rotifer species showing significant amounts of unexplained mortality was therefore studied by correlation procedures.

The analysis revealed no significant relationships between the two sets of variables (Table 12–3). The evidence thus indicates that the large herbivores do not play any role in determining the unexplained mortality of rotifers.

In summary, the three separate attempts to explain the mortality of rotifers which cannot be explained by *Chaoborus* predation failed to suggest other mechanisms by which mortality might occur except in connection with food quality. It does appear that the mortality of *Conochiloides,* the most abundant rotifer species, is directly related to the presence of large amounts of algal biomass belonging to the Chlorophyta, Dinophyceae, or Cryptophyceae. Comparable evidence for *Tetramastix* is not strong. The data on relative mortality rates of *Tetramastix* are of considerably lower quality than the data on relative mortality rate of *Chaoborus,* however, as *Tetramastix* is much less abundant. Thus the failure to demonstrate such a strong relationship for the mortality of *Tetramastix* could in part be accounted for by the much greater amount of noise in the data for this species.

In the comparison of species similarities in Chapter 10 (Table 10–3), the rotifers showed a tendency to segregate from other herbivores when growth and mortality were the basis for comparison (ΔN matrix). It is likely that the distinctive (nonpredation) mortality control mechanisms in rotifers, such as the apparent nutritional mechanism indicated for *Conochiloides* in this analysis, are in part responsible for this segregation.

Chapter 13

Community Trends in Mortality

Given the information which has been developed in the preceding chapters, it is possible to test some general hypotheses relating mortality rates to measurable properties of zooplankton species. Mortality statistics available for testing include (1) absolute total mortality (individuals per day), (2) relative total mortality (percentage/day), and (3) relative predation mortality (percentage/day). The one quantifiable property common to all developmental stages and species is size, which can be expressed either in terms of wet weight or length. The general hypothesis to be tested here is that definite statistical relationships exist between the various measurements of mortality rate and the size of herbivores.

Prior to the analysis, some decision must be made about the inclusion of the rarest species in the analysis. The discussion of errors in mortality rate estimates would suggest that inclusion of the rarest species could be misleading, and that the relative mortality rates are more sensitive to distortion than the absolute ones. For these reasons, the rare rotifers with indistinguishable eggs are eliminated from the analysis of absolute mortality (*Hexarthra, Polyarthra, Trichocerca*), and the categories with substantial numbers of weeks below the 10% error criterion are additionally eliminated in the relative mortality analysis and predation mortality analysis (the three rotifers already named plus *Keratella, Bosmina, Moina*). The broad conclusions would be qualitatively the same with or without these exclusions, however.

1. Absolute Total Mortality versus Size

The copepods are treated as species rather than as stages, as the analysis attempts to identify trends across species. A Spearman Rank Correlation analysis shows that absolute mortality is related to average wet weight per individual ($r_s = 0.64$, $P < 0.05$). Since average total mortality must be closely balanced with average total production, the trend implies a corollary significant relation between production and size. The significance of the trend is best discussed in terms of production and will thus be deferred to the analysis of production trends in the community.

2. Relative Total Mortality versus Size

The relationship between relative mortality and length or wet weight is shown in Figure 13–1. For copepods, the major developmental stages are separated so that the variable operation of mortality mechanisms within species can be included in the analysis. The figure is constructed on the basis of mortality vs length, but it is also possible to show wet weight on the same graph with a reasonable degree of accuracy because of the close relation between length and weight. The line indicated in Figure 13–1 is the best

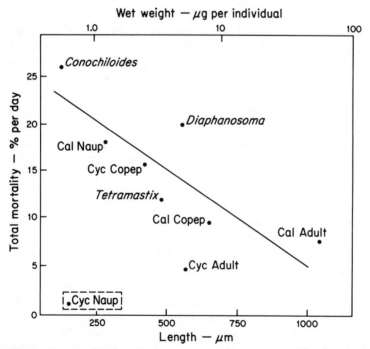

Figure 13–1. Relationship between the total relative mortality (weighted mean) of zooplankton herbivores in Lake Lanao and their length (μm) or wet weight (μg, wet).

linear fit to the mortality–length relationship as follows:

$$Y = 24.1 - 0.019X$$

where Y is the mean total relative mortality (percentage/day) and X is the mean length of the organism (μm). The relationship is significant ($P = 0.02$). Length explains about 55% of the variance in relative mortality. The analysis excludes cyclopoid nauplii, which show a mortality highly divergent from the trend for unique reasons explained below.

The relationship between total relative mortality and wet weight is similar to the total relative mortality–length relationship, but is not quite so strong ($P = 0.10$). The equation is:

$$Y = 16.5 - 0.33X$$

where Y is identical to the previous equation and X is the wet weight of an average individual (μg).

Figure 13–1 and the accompanying statistical analysis demonstrate a pervasive and significant trend toward higher average mortalities in smaller species and developmental stages. While factors other than size are unquestionably significant and contribute to the scatter of points, the existence of a simple relationship between size and mortality in a natural community over an extended period of time has important implications for its organization and function.

3. Predation Mortality versus Size

The last comparison involves predation mortality and the length and weight variables that have already been used in the other comparisons. The relation depicted in Figure 13–2, which is statistically significant ($P = 0.05$), is as follows:

$$Y = -4.00 + 0.033X$$

where Y is the loss to predation (percentage/day) and X is the length of an average individual (μm). The length of herbivores explains about 45% of the variance in their relative loss to predators. As in Figure 13–1, Figure 13–2 shows the scale of wet weights on the abscissa as well as the scale of lengths, although the locations of the points themselves are based on length. The relationship to weight is also significant ($P = 0.04$) and the line of best fit is:

$$Y = 2.50 + 2.39X$$

where Y is as in the previous equation and X is the weight of an average individual (μg). The trend is remarkably uniform within the copepods but the rotifers and Cladocera follow the trend much less closely. Although size is clearly important, the identity of the species is obviously important as well (cf. Lewis, 1977a).

The significance of a general relationship between body size and predation

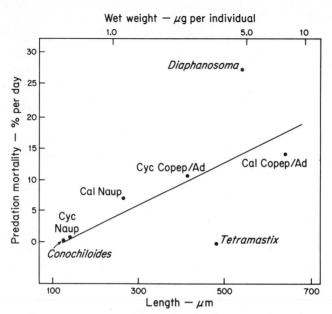

Figure 13–2. Relationship between predation mortality of zooplankton herbivores in Lake Lanao and their length (μm) or wet weight (μg).

mortality in the context of plankton community structure has received a great deal of recent attention (reviewed by Brooks, 1968; Zaret, 1975; Hall et al., 1976; Macan, 1977). Several kinds of empirical evidence suggest that larger zooplankton organisms (or organisms of larger apparent size: O'Brien et al, 1976) are subject to substantially greater predation pressures from vertebrate planktivores. Some evidence also suggests that invertebrate predators have the opposite effect (reviewed by Hall et al., 1976), but the supporting data are less comprehensive.

The potential basis for community order related partly to size in plankton systems is well illustrated by Allan's (1976) analysis of life history features as elements of adaptive strategies in zooplankton. Allan concludes, largely on grounds of capacity for increase, that the order of opportunism among zooplankton should be rotifers > Cladocera > copepods. As a corollary to this, he reasons that predation stress will follow the same order. In Lake Lanao the total relative mortality follows roughly this order, although developmental stages within species may vary enormously (Fig. 13–1). The mortality trend is not, however, due to higher predation stress on smaller organisms. In fact the opposite is true (Fig. 13–2). The major taxa may still be ranked as suggested by Allan with respect to opportunism, but factors other than predation must enforce this ranking, at least in Lanao. The identity of some of these factors will become evident in the analysis of growth control mechanisms.

The total predation losses of a relatively complex community of zooplankton organisms have seldom been quantified over any considerable period of time, so comparisons must be made with caution. Figure 13–2 runs counter to the general notion that increased size results in a significantly lower risk of predation from invertebrate planktivores, but the data are taken from a community in dynamic equilibrium with regard to species composition, whereas much of the theory relating predation loss to body size applies to nonequilibrium conditions, i.e., it explains how species composition is established but does not necessarily explain the relative abundance of species at equilibrium. It is also clear from Figure 13–2 that factors other than size have considerable influence, as indicated by the scatter of points and the uniquely low total mortality of cyclopoid nauplii.

Because the cyclopoid nauplius is outside the general trend of total mortality in the community, it is of special interest. Cyclopoid nauplii obviously suffer very low mortality from *Chaoborus* predation even though cyclopoid nauplii are extremely abundant. *Chaoborus* is a lurking predator whose success may be markedly affected by the average movement rate of the prey organisms (Gerritsen and Strickler, 1977). Gerritsen (1978) has recently shown that the losses of varous cyclopoid stages to *Chaoborus* vary greatly because of differences in swimming patterns. Cyclopoid nauplii move very little and are consequently less likely to encounter a predator or to attract its attention. This confirms earlier speculation that the very low electivity of *Chaoborus* for nauplii may be due to special escape or cryptic behavioral mechanisms not shared by copepodids and adults (Lewis, 1977a).

Figures 13–1 and 13–2 show that the low vulnerability of cyclopoid nauplii to predation does not really provide the entire explanation for uniqueness of cyclopoid nauplii with regard to community trends in total relative mortality. The rotifer *Conochiloides* also experiences very low predation losses, and yet experiences sizable total mortality, as would be expected from the community trends. The failure of cyclopoid nauplii to fit the community trends derives from very low total mortality, which is only partly explained by its immunity from predation. The contrast in total relative mortality between cyclopoid nauplii and *Conochiloides* is explained very simply by a critical biological difference between the two. Cyclopoid nauplii never reach maturity (they merely become copepodids) and thus are not subject to mortality mechanisms which specifically affect mature organisms. *Conochiloides*, on the other hand, does mature at a small size and in the absence of significant predation experiences mortality from other sources. The divergence of the cyclopoid nauplii from the trend in Figure 13–2 is thus understandable. Calanoid nauplii, although equally immune to mortality affecting mature organisms, are subject to significant predation and thus fit the community pattern, unlike cyclopoid nauplii. It is the combined immunity from predation and senescence that gives the cyclopoid nauplii their special position in Figure 13–2.

The analysis has shown that both total mortality and predation mortality are significantly related to herbivore size in Lake Lanao, but the trends run in opposite directions. Although there is a steep increase in predation losses with body size, the increase is more than compensated by the existence of mechanisms of mortality other than predation which take an exceptionally heavy toll of the smallest organisms.

The production of the smallest herbivores, in which predation is a minor loss, for the most part follows two pathways. If energy is incorporated in a small herbivore which is incapable of growing large (rotifer), then the energy will flow according to nonpredation mortality mechanisms, which will lead to microbial–detrital channels. If the energy is incorporated in a small herbivore capable of growing large (cyclopoid nauplius), then mortality is minimal until sufficient growth occurs to bring the body size into the range of significant predation. These patterns are illustrated in Figure 13–3.

The complete diversion of some energy away from primary carnivores and toward the microbial–detrital channel would seem to provide some evidence of inefficiency in the primary carnivores. This impression is somewhat misleading because of the small amount of production attributable to the

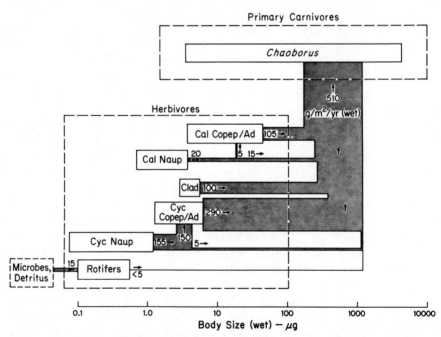

Figure 13–3. Fate of herbivore secondary production in all of the herbivore categories of Lake Lanao. Note that production of the smallest herbivores either flows away from the primary carnivores channels into microbial–detrital channels or is shunted into larger herbivore size categories before flowing to the primary carnivores.

herbivores whose production flows into microbial–detrital channels (Fig. 13–3). The entire rotifer component accounts for only 3.5% of herbivore production (Table 9–1), hence the failure of primary carnivores to exploit this component fully is not very significant from the energetics viewpoint. The production of small nauplii, which also avoid predation, eventually is shunted to the primary carnivores anyway, as the harvesting is merely deferred until the organisms are larger (Fig. 13–3). Thus the primary carnivores do intercept about 95% of herbivore production even though they do not prey effectively on the smallest herbivores.

Chapter 14

Control of Zooplankton Production by Food Quantity and Quality

Analytical Methods: Herbivores

Zooplankton growth and reproduction, which are measured here in terms of secondary production, are potentially affected by quality and quantity of the food resource. The following analysis attempts to identify and quantify this influence on Lake Lanao herbivores. Mortality mechanisms are potentially confounding to the analysis because mortality exercises some feedback control over total growth and reproduction through its alteration of the numerical base to which growth rates apply. It would seem that the influence of mortality on secondary production (P) could be minimized by expressing P on a relative basis (P/B). Although this method of expressing production would reduce the influence of the numerical base, it would not be defensible because of the manner in which P is estimated. The estimate of secondary production is dependent partly upon the number of individuals and partly upon their age distribution. Since the ratio of P to B is dependent on the age structure of a population as well as the conditions for growth, changes in age structure might cause substantial alterations in the ratio of P to B even when there is no change in growth control mechanisms (cf. Edmondson, 1965). The analysis of growth control mechanisms must therefore be based on P rather than P/B. The potential confusing effect of mortality on the analysis can be reduced, however, by examination of lag effects. Time lags allow for compensatory growth to alter the numerical base in response to changes in growth control mechanisms.

The relationship of zooplankton production to the phytoplankton food resource is analyzed here in stages from the following variable sets: (1) abundance of five major phytoplankton taxonomic groups, (2) abundance of six major phytoplankton size groups, and (3) abundance of 50 individual phytoplankton species. The analyses are based on correlation matrices derived from the production of the herbivore categories over successive weeks and the concurrent abundances of phytoplankton species or phytoplankton groups. It is important to note that this analytical approach, while based on biomass of food, is sensitive to the *variation* in biomass of food in relation to *variation* in herbivore production, and not to the mean values for these variables. The basic assumption is that variation in a critical food resource, regardless of its mean availability, should be manifest as corresponding variation in the consumers.

As explained in connection with the production computations, the production of a given herbivore category over week k was obtained from the average demographic characteristics of the population at the beginning and the end of the week, or from a numerical model of population change assuming a linear alteration in demographic features between the beginning and the end of the week. The basis for the present analysis is the zooplankton production over week k and the phytoplankton abundance over the time period immediately preceding this (week $k-1$). Four kinds of correlation matrix are used: (1) Pearson Product–Moment correlation coefficient matrix on untransformed variables, (2) Pearson Product–Moment correlation coefficient matrix on log-transformed variables, (3) Spearman Rank correlation coefficient matrix, and (4) Pearson Product–Moment correlation coefficient matrices incorporating different time lags.

Log transformation of variables and Spearman Rank correlation always yielded essentially identical interpretations, but these differed from interpretations that would be obtained from a Pearson Product–Moment correlation coefficient matrix on untransformed variables. Examination of the underlying frequency distributions showed that conclusions drawn from untransformed variables would be unreliable as a result of a small number of outlying points which distort the frequency distributions. For this reason, the log-transformed variables were used throughout the analysis. A 1-week lag is inherent in the correlations to allow the numerical base affecting P to respond to algal food quality. Examination of the correlation coefficient matrices incorporating additional time lags of 1 or 2 weeks or a negative 1-week lag that would make the food and production data exactly coincident showed that the interpretation would be essentially the same for no additional lag, for a 1-week additional lag, or for a negative 1-week lag, but that the number of significant correlations and intensity of correlations would be reduced with a 2-week lag. The data analysis is therefore based on correlation of production week k with phytoplankton abundance week $k-1$ as initially proposed.

An important distinction between ingestion and nutrition must be made before the results are evaluated. Correlations between secondary production and food abundances are affected by factors other than the ingestion rate of the food by the herbivore. Ingestion of substantial quantities of a given phytoplankton species may not result in a higher secondary production if the phytoplankton species is not digestible or nutritious. Thus the correlations test for the total nutritional effectiveness of phytoplankton species, including ingestion, digestion, and nutritive value.

Importance of Major Phytoplankton Taxonomic Groups as Herbivore Foods

Figure 14–1 shows the results of correlation analysis comparing the production of zooplankton herbivores and the abundance (measured as biomass) of the five major groups of phytoplankton that could serve as food sources for herbivores. All of the relationships indicated in Figure 14–1 are significant at $P < 0.01$.

Interpretation of Figure 14–1 must be made with caution, as each of the major taxonomic groups of phytoplankton is composed of a number of species which may differ in usefulness to herbivores. In addition, an association between herbivore production and the abundance of a particular phytoplankton group provides only circumstantial evidence that the phytoplankton group is important to the nutrition of the herbivore. The problem of spurious correlation will be considered briefly here and in greater detail subsequently with the analysis of individual species.

Cyclopoid production shows strong evidence of association with abundance of diatoms and of bluegreen algae. As the abundances of these two major food sources are not correlated with each other (Lewis, 1978b), the correlations must be explained by separate affinities of the zooplankton herbivores with the diatoms and bluegreen algae. The results thus indicate that one or more of the major species in each phytoplankton group is an important resource for all of the developmental stages of cyclopoid copepods. The diatom flora is strongly dominated by *Nitzschia baccata,* which is therefore likely to be important in explaining the relationship of diatom biomass to cyclopoid production. The bluegreen algae are represented in Lake Lanao by several filamentous and coccoid species of high abundance, and Figure 14–1 does not indicate which of these would be most important in explaining the association with cyclopoid production. The adult cyclopoids also show a significant association with cryptomonads that is not shared by the younger developmental stages.

The calanoids do not show the general affinity for bluegreen algae that is characteristic of the cyclopoids, but do share the affinity for the diatoms. The diagram thus suggests that the cyclopoids use a broader range of food resources than the calanoids. The cladocerans all show a remarkable affinity

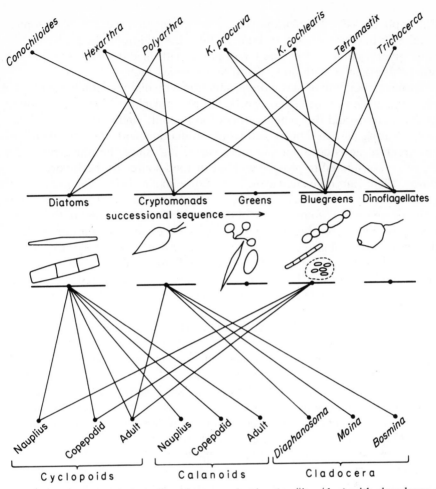

Figure 14–1. Association of herbivore production (μg/liter/day) with abundance of five major phytoplankton taxonomic groups (biomass). Each line joining an herbivore category with a phytoplankton taxon indicates a highly significant positive correlation between herbivore production and phytoplankton ($P < 0.01$).

for the cryptomonads and no significant affinity for the diatoms or bluegreen algae.

Rotifer production is associated with four of the five major phytoplankton groups, but individual species differ considerably in their associations. The affinities are particularly strong with the bluegreen algae and secondarily with the cryptomonads and dinoflagellates. The diatoms play a much smaller role than with the copepods.

The average percentage biomass composition of the Lake Lanao phytoplankton, as given by Lewis (1978a), is as follows: Cyanophyta, 19%;

Chlorophyta, 34%; Bacillariophyceae, 37%; Dinophyceae, 2%; Cryptophyceae, 8%. In terms of species composition, the Chlorophyta are by far the most diverse, followed by the Cyanophyta with medium diversity and the other three groups with only two to four species each.

Strong association of herbivore production with diatom abundance is to be expected in view of the high proportion of phytoplankton biomass accounted for by diatoms and the generally high food quality and digestibility of diatoms. Strong association of herbivore production with the abundance of bluegreen algae is considerably more surprising, since the bluegreen algae are often considered to provide an inferior food source for herbivores (e.g., Burns, 1968; Hargrave and Geen, 1970; Bosselmann, 1975). The use of bluegreen algae by herbivores would not really be inconsistent with the literature, however, as some zooplankton populations are known to subsist largely or entirely on bluegreen algal foods. The most relevant example in this case is the subsistence of *Thermocyclops hyalinus* in Lake George on *Microcystis* (Moriarty et al., 1973). Infante (1978) has also shown that cyclopoids and calanoids in Lake Valencia, Venezuela, ingest and digest large amounts of *Lyngbya limnetica,* which also occurs in Lake Lanao. Since the peak abundances of bluegreen algae are entirely out of phase with peak abundances of diatoms and cryptomonads (Lewis, 1978b), it would be unlikely that an association of major zooplankton herbivore production with the abundance of bluegreen algae could occur by spurious correlation through some other variable. Occurrence of high herbivore production in connection with the high levels of bluegreen algal biomass could scarcely be explained by a mechanism other than the use of bluegreen algal biomass as food, since significant amounts of alternative food sources are not available when bluegreen algae occur in great numbers.

The complete absence of significant relationships between herbivore production and abundance of green algae (Fig. 14–1) is quite unexpected in view of the large contribution of green algae to the total algal biomass. It will become clear later in the analysis that some of the green algal species are in fact suitable as herbivore food, but that most are unsuitable, and the resulting mixture of contributions from suitable and unsuitable species to the total pool of green algal biomass prevents any general correlation between herbivore production and the abundance of green algae.

The cryptomonads show an unexpectedly large number of significant associations with herbivore production in view of their low species diversity and generally low biomass. The cryptomonads do become very significant elements of the phytoplankton community at certain times of the year, and are evidently very good food sources for certain herbivores, notably the Cladocera and some rotifers. Specialization of certain rotifers on cryptomonads is documented in the literature (e.g., *Polyarthra,* Edmondson, 1965; Pourriot, 1965; Fairchild et al., 1977; *Keratella,* Nauwerck, 1963).

The dinoflagellates are even less abundant than the cryptomonads, but nevertheless show significant associations with some of the rotifer species. Rotifers are usually assumed to feed for the most part on very small particles (Edmondson, 1965; Hutchinson, 1967). Recent evidence indicates, however, that rotifers may use different feeding mechanisms to accommodate for variation in food particle size and shape (Gilbert and Starkweather, 1977; Starkweather and Gilbert, 1977). A general affinity for flagellates as food is apparent in the rotifer literature, and small planktonic rotifers are known to be able to capture relatively large flagellates (e.g., Pourriot, 1963). Green et al. (1976) record the ingestion of *Peridinium* by *Brachionus* in an Indonesian lake. Since the dinoflagellates show significant growth only under very extreme conditions (Lewis, 1978b), spurious associations are extremely unlikely except possibly through some of the bluegreen species. These questions will be considered in greater detail in connection with the analysis of individual species.

The relationships depicted in Figure 14–1 indicate that the successional sequence of phytoplankton taxa (Diatom/cryptomonad → Green → Bluegreen → Dinoflagellate; Fig. 3–2) will not be translated into simple zooplankton successional sequences. Dependence of production in a given herbivore species on the abundance of algal taxa not adjacent to each other in the successional sequence greatly increases the possible variety in herbivore sequences. This is compounded by the greater inertia of herbivore populations, particularly those which show extended development. Furthermore, it is clear from the mortality analysis that primary carnivores exert strong control over herbivore composition and abundance but are essentially independent of the phytoplankton sequence. The temporal ordering of phytoplankton taxa on environmental gradients (Lewis, 1978b) is thus not reflected in any simple way by herbivore populations.

Bacteria and detritus are potential supplementary foods for herbivores, but are not likely to be quantitatively important in the Lake Lanao system. Detritus is present in extremely small amounts ($< 5\%$ of total particles). This is partly explained by the large size and great depth of the lake, which minimize the transfer of sedimented material and allochthonous material into the plankton environment. In addition, large tropical lakes generally may have smaller amounts of nonliving organic particulates in the plankton zone because of the rapid degradation of organic particulates at high temperatures (Lewis, 1974).

Bacteria counts with oil immersion phase optics showed an annual average of about 800,000 bacteria cells per ml in Lake Lanao. These were mostly very small, with an average volume of about $0.1 \ \mu m^3$/cell. Such small average sizes for bacterial cells in freshwater systems are evidently not unusual. Hobbie and Rublee (1975), for example, used an average carbon content per cell for bacteria of an arctic pond which would correspond to a cell volume of about $0.3 \ \mu m^3$. Possibly many of the very small cells are metabolically inactive (J. Hobbie, personal communication). Even assuming all are active, and allowing for the standard increase in QO_2 with cell size, which would indicate metabolic rates 26 times higher than for Lanao phytoplankton (Lewis, 1974, Table 11), the bacteria of Lake Lanao would be expected to show a maximum production of only about 5% that of the phytoplankton. Even this maximum figure would not sustain herbivore production. Thus if bacteria are an herbivore food source in the Lanao system, they are probably very secondary to the phytoplankton.

Importance of Different Phytoplankton Size Fractions as Herbivore Foods

Filter-feeding zooplankton have the ability to select certain size fractions and reject others (e.g., Wilson, 1973; Berman and Richman, 1974; Friedman and Strickler, 1975), and in some cases are constrained to the filtration of particles in a certain size range by the physical nature of their filtration apparatus. Raptorially feeding herbivores are limited by similar constraints but for different reasons. The Lake Lanao zooplankton community is principally composed of filter-feeding species, but the copepods are probably capable of switching between filter feeding and raptorial feeding (McQueen, 1970; Lam and Frost, 1976), as are some of the rotifers (Hutchinson, 1967).

The hypothesis to be tested here is that herbivore production is associated with the abundance of phytoplankton particles of specified size. As in the analysis of mortality mechanisms in a previous chapter, the phytoplankton community is divided into six size categories, each containing approximately 11 phytoplankton species. The numerical abundance of phytoplankton particles in each of these categories is then determined by summation of the abundances of the species within the group. The abundances of phytoplankton size groups expressed in this manner are then compared with the productivities of the 16 herbivore categories.

The analysis is repeated three times. The first time, the separation of phytoplankton species is made on the basis of the average volume of a plankton unit (individual) for each of the 70 species. The ranges of volumes for the size categories are shown in Table 14–1. The second analysis is based on the greatest axial linear dimension (GALD) of average plankton units for each of the 70 species. The size ranges are indicated in Table 14–1. The third analysis is based on the second greatest axial linear dimension (SGALD), the importance of which might derive from the effect of width on food particle ingestibility.

The analysis based on biomass (i.e., volume) of average plankton units appears first in Table 14–1. All herbivore categories show significant relationships with one or more phytoplankton size fractions. Close examination of the results shows, however, that there is little meaningful pattern to the sig-

Table 14-1. Summary of Correlation Data Between Zooplankton Secondary Production in 16 Herbivore Categories and the Abundance of Various Size Fractions of Phytoplankton[a]

Phytoplankton category	Cyclopoid			Calanoid					Rotifers					Cladocera		
	N	C	A	N	C	A	CD	HI	PV	KP	KC	TO	TB	DM	MM	BF
Size—μm³																
2–18					+		+							+		
19–60	+	+	+				+									
61–120		+	+		+			+	+	+		+	+	+		+
121–320	+			+						+			+		+	+
321–800			+			+			+		+					
801–50000																
Size—GALD, μm																
1.0–6.5					+		+		+					+		+
6.5–12.5					+									+	+	+
12.6–22								+						+		
23–39			+	+		+	+			+	+			+		
40–65	+	+	+				+			+	+	+	+			
66–500	+	+	+	+		+			+			+	+			
Size—SGALD, μm																
0.5–2.2	+	+	+	+	+	+	+			+		+	+	+		
2.3–4.5	+	+	+				+			+			+	+		
4.6–6.5				+	+		+							+		
6.6–11.5							+							+		
11.6–27									+							
28–300											+					

[a] The symbol + indicates a positive correlation (log–log Pearson Product–Moment) significant at $P \leq 0.01$. Phytoplankton are first grouped by biomass (μm³ per individual), then by greatest axial linear dimension (GALD, μm), then by second greatest axial linear dimension (SGALD, μm).

nificant relationships. For example, phytoplankton size fractions 2, 4, and 6 are associated with production of cyclopoid adults, whereas fractions intermediate in size between these are not. Results of this type would strain any interpretation based on a strictly physical separation of particles by volume. It will become clear in the analysis of individual species that the correlations are in fact mostly accounted for by the chance distribution of important phytoplankton species between the various volume fractions.

The volume of a plankton unit might be considered a poor basis for analysis because herbivores are likely to be more sensitive to the dimensions of a particle than to its volume. Although the dimensions of a particle could be correlated with its volume, the shape of phytoplankton organisms varies enormously, which would partly mask any such correlation. The relationship between herbivore production and phytoplankton size categories is therefore reconsidered on the basis of greatest axial linear dimension (GALD), as shown in the second portion of Table 14–1. The results for GALD are more coherent than the results based on particle volume. The cyclopoids show a definite affinity for plankton units with very high GALD values. Such plankton units are usually filaments (e.g., *Lyngbya, Anabaena*) or elongate unicells (e.g., *Nitzschia*). Since the cyclopoids account for over half of the total herbivore secondary production (Table 9–1), this is a very important feature of herbivore grazing. A completely different analysis based on phytoplankton demographic data confirms that filaments are particularly vulnerable to grazing in Lake Lanao (Lewis, 1977c).

The size analysis based on GALD also indicates that the Cladocera, in contrast to cyclopoids, prefer particles with low to moderate GALD values. No overall trends are evident in the rotifers, which appear to prefer particles with a wide range of GALD values, nor in the calanoids, which show scattered affinities. Some individual developmental stages and species from among the calanoids or rotifers do show evidence of affiliation with particular GALD size ranges, however. Calanoid copepodid production shows strong association with particles of low GALD, *Trichocerca* production with particles of moderate GALD, and *Keratella* spp. with particles of high GALD. Others show only scattered relationships.

The final analysis is based on second greatest axial linear dimension (SGALD), which might be particularly relevant to the ingestion of particles with elongate shapes. The results compare favorably in coherence with the results for GALD. The cyclopoids show a strong affinity for minimum SGALD values, indicating in view of the GALD results a preference for slim, elongate particles. This in turn fits with the taxonomic analysis of foods suggesting a dependence of the cyclopoids on the diatoms and bluegreens, many of whose species are elongate filaments or unicells. The calanoids show some evidence of association with low SGALD. The rotifers show no universal tendencies. *Conochiloides, K. procurva, Tetramastix, and Trichocerca* show association with low SGALD to varying degrees of certainty, while *Polyarthra* and *K. cochlearis* show the opposite tendency. Production of the dominant cladoceran (*Diaphanosoma*) is associated with particles of moderate SGALD, suggesting together with the GALD data a preference for small to medium-size particles of round or oblong shape.

In general, the analysis suggests that size is an important quality of foods, especially in certain herbivore categories, but that factors other than size must be important in determining food quality and thus secondary production in the herbivore categories. Where size is important, it is clear that GALD and SGALD are more critical than the particle volume.

Importance of Individual Phytoplankton Species as Herbivore Foods

The entire correlation matrix was obtained for production of the 16 herbivore categories and abundance of the 50 most abundant phytoplankton species. For each herbivore category, there were 50 possible correlations, of which approximately two to three should be significant and positive by chance alone. The number of significant positive correlations in each herbivore category averaged 11, or considerably above the number expected by chance. Since the abundances of phytoplankton species are in some cases correlated with each other, however, it is necessary to treat the data further by multivariate techniques to determine which of the phytoplankton species actually account for the bulk of variance in herbivore productivity. All phytoplankton species showing significant positive correlations were thus subjected to step-wise multiple regression with production in a particular herbivore category. The list of phytoplankton species was in this way typically reduced to a few species (two to five) which make significant separate contributions to the variance of herbivore production. The amount of variance accounted for by each of the food species was then tabulated and totalled as shown in Table 14–2.

The phytoplankton species which show significant relationships to herbivore production are not randomly distributed among the major taxa listed in Table 14–2. The bluegreen algae, cryptomonads, diatoms, and dinoflagellates all show a substantial percentage of significant relationships to herbivore production, whereas the green algae show a much lower percentage. This accounts for the failure of the green algae as a whole to show significant relationships to herbivore production at the division level, as indicated previously. Although certain species of green algae (*Oocystis lacustris, Franceia droescheri*) show every indication of being important food items, similar closely related species sometimes do not. The failure of very many common green algal species to show any significant relationships at all to herbivore production is an outstanding feature of the results and is probably

Table 14-2. Variance in the Production of 16 Herbivore Categories Accounted for by the Abundance of the 50 Most Common Phytoplankton Species in Lake Lanao[a]

	Cyclopoids			Calanoids			Rotifers							Cladocera			Total
	N	C	A	N	C	A	CD	HI	PV	KC	KP	TO	TS	DM	MM	BF	
Cyanophyta																	
Chroococcus minutus																	0
Aphanothece nidulans			20				27						12				59
Aphanocapsa elachista																	0
Gloeothece linearis							7										7
Dactylococcopsis fascicularis																9	9
Dactylococcopsis wolterecki																5	5
Anabaena sphaerica													20				20
Anabaena spiroides		25	3							21							49
Lyngbya limnetica	4										11						15
Rhabdoderma sp. 1		4															4
Synechococcus sp. 1																	0
Euglenophyta																	
Trachelomonas bacillifera																	0
Trachelomonas sp. 2																	0
Chlorophyta																	
Chlamydomonas sp. 1																	0
Tetraedron minimum																	0
Sphaerocystis schroeteri																	0
Chlorella sp. 1						4											4
Oocystis submarina																	0
Oocystis lacustris	20			15	33									8			76
Franceia droescheri				7	8	9								38			62
Chodatella subsalsa																	0

														Total
Chodatella sp. 1							6							6
Ankistrodesmus setigerus											5			5
Ankistrodesmus gelifactum							4							4
Kirchneriella elongata														0
Kirchneriella obesa														0
Selenastrum minutum														0
Selanastrum sp. 1							19					15	30	64
Dictyosphaerium pulchellum														0
Dimorphococcus lunatus														0
Coelastrum cambricum														0
Scenedesmus quadricauda														0
Scenedesmus ecornis														0
Scenedesmus sp. 1														0
Crucigenia rectangularis														0
Coccomyxa sp. 1					3									3
Closterium sp. 1														0
Closterium sp. 2								10				30		40
Staurastrum paradoxum						4								4
Unknown 28														0
Unknown 29														0
Unknown 36														0
Chrysophyceae														
Chromulina sp. 1														0
Bacillariophyceae														
Nitzschia baccata	26	8	27	40	12	40	11	8						172
Melosira granulata														0
Melosira agassizii			2											2
Dinophyceae														
Gymnodinium sp. 1							30			7				37
Peridinium sp. 1									39					39

Table 14–2. (continued)

	Cyclopoids			Calanoids				Rotifers						Cladocera			
	N	C	A	N	C	A	CD	HI	PV	KC	KP	TO	TS	DM	MM	BF	Total
Cryptophyceae																	
Rhodomonas minuta	6	9							38					17			70
Cryptomonas marssonii			14					4	15			19				18	41
Total variance accounted for (%)	56	46	66	62	56	57	34	63	64	39	50	26	32	63	50	52	

a Under each herbivore category the percentage of variance accounted for is indicated for those phytoplankton species making a statistically significant ($P < 0.05$) contribution to variance in production, as determined by step-wise multiple regression. See text for details.

of great ecological significance. At least some of the variation between foods is probably related to differences in digestibility (Naumann, 1923; Fryer, 1957b; Porter, 1973, 1975), but other factors may be involved as well.

Totalling across all herbivore species in Table 14–2 yields a statistic which reflects the general importance of each of the phytoplankton species in accounting for variance in herbivore production. Twenty-six of the 50 species show a total of 0. The frequency distribution of totals is skewed to the right and is significantly different from the expected normal distribution of values obtained by chance (Komogorov–Smirnov criteria, $P < 0.05$). The results thus indicate that an unexpectedly large number of phytoplankton species bear no relation whatever to the trends in herbivore production. To put this another way, the phytoplankton species which show significant relationships to herbivore production are fewer in number than would be expected to result from the random distribution of significant relationships in Table 14–2. Certain phytoplankton species appear to be of extraordinary importance in regulating herbivore production, while many other species are of trivial importance.

The data of Table 14–2 can be used to test hypotheses relating the importance of a phytoplankton species in accounting for herbivore production to its morphological or demographic properties. The cross-column total of percentage was used as a relative indicator of the importance of a given phytoplankton species as an herbivore resource. It should be noted that the theoretical range of this total is 0 to 1600 rather than 0 to 100 because the percentages for all 16 herbivore categories are summed to obtain the total.

The total percentage of variance accounted for by each of the 50 phytoplankton species was compared by correlation analysis with measurements of the average abundance and average size of the phytoplankton species. The variables were log transformed. Results are given in Table 14–3.

Table 14–3 shows that the overall importance of a phytoplankton species in accounting for variation in herbivore production is related to abundance

Table 14–3. Correlation between Phytoplankton Species Attributes and the Total Amount of Variance in Herbivore Production Accounted for across All 16 Herbivore Categories

	Phytoplankton properties				
	Abundance		Size		
	μm^3/ml	Plank units/ml	μm^3	GALD—μm	SGALD—μm
Total percentage variance	0.30[a]	0.35[b]	−0.10	0.05	−0.33[b]

[a] $P < 0.05$.
[b] $P < 0.01$.

measured either in terms of biomass or in terms of numbers of plankton units (i.e., individuals) per unit volume of water. Abundant species are more likely to be important determinants of phytoplankton variation. Although the relationship is statistically significant, the correlation coefficient itself is not impressively high, implying that abundance is only one of several important factors which affect the overall importance of a phytoplankton species as a determinant of herbivore production.

Table 14–3 also shows a relationship between total percentage of variance accounted for and the size of average plankton units for all 50 phytoplankton species. The table shows that the longest dimension (GALD) of a phytoplankton organism is not a satisfactory universal index of its usefulness as an herbivore food. This is to be expected in view of the foregoing analysis of individual herbivore categories, which showed an affinity for high GALD in copepods, for low to moderate GALD in Cladocera, and no fixed affinity in rotifers. Table 14–3 shows, however, that the SGALD is a significant universal index of the overall nutritional worth of phytoplankton, and that the secondary dimensions should be small if the food is to be used in large amounts by herbivores. Narrow filaments and unicells as well as small coccoid forms would meet these requirements, despite their dissimilarity in overall shape.

The total variance in herbivore production accounted for in Table 14–2 by individual phytoplankton species is surprisingly high (average, 51%) and indicates an intimate dependence of the production of individual herbivore categories on a limited number of ingestible, nutritionally sufficient phytoplankton species. The accompanying size and taxonomic analysis indicates that these species may come from any one of the major taxonomic groups, that they are likely to be reasonably abundant, and that they are likely to have small cross-sectional measurements. The total variance accounted for is lowest for some of the rotifer species. As with the mortality analysis, the controlling factors are more in doubt for the rotifers than for the copepods or Cladocera.

Importance of Food in Determining Primary Carnivore Production

The data on *Chaoborus* production can be subjected to the same kind of analysis which has already been completed for the herbivores. The analysis is more straightforward because the exact food habits of *Chaoborus* are known. The role of food supply in regulating *Chaoborus* production can therefore be established without exploratory analysis to determine which food species are important.

Important food organisms for the Lake Lanao *Chaoborus* population include cyclopoid copepodid/adults, calanoid copepodid/adults, *Keratella cochlearis,* and the three cladoceran species (Lewis, 1977a). A composite food abundance was obtained by totalling the biomass of these categories for each week of the study period. This composite food abundance was then compared by correlation techniques with *Chaoborus* production data to determine the dependence of *Chaoborus* production on total food availability.

In the first comparison, *Chaoborus* production (μg/liter/day) was compared with average food abundance over the preceding 7-day period (μg/liter). The correlation was significant ($P < 0.05$) but accounted for only a small proportion of total variance ($r = 0.16$, $r^2 = 0.07$). The analysis was repeated after log transformation of both variables, which appeared to be justified on the basis of poor distribution of points prior to transformation. The resulting relationship was not significant at $P = 0.05$. Likewise, incorporation of lags equal to 1 or 2 weeks between the variables did not result in improvement of the relationship. The analysis was also repeated using only the Cladocera, since the electivity values for cladocerans are very high (Lewis, 1977a). The resulting relationship was not significant at $P = 0.05$.

It would be possible to conduct a more rigorous analysis by determining the relationship between food abundance and productivity of separate size groups or instars of *Chaoborus,* thus accounting for some shift in feeding preference with size. This does not appear to be warranted, however, in view of the poor overall relationship between food abundance and *Chaoborus* production. The instars overlap a great deal in food preference and any strong relationship between food abundance and *Chaoborus* production should be evident in the pooled data.

The analysis demonstrates that week to week variation in *Chaoborus* production is much less under the control of food supply than is the production of herbivores. Since most of the variance in *Chaoborus* secondary production cannot be accounted for by food availability, other mechanisms must account for the bulk of variation. Although mortality mechanisms have already been dealt with for herbivores, *Chaoborus* mortality was not partitioned between predation and other mechanisms because the predation rates of fish on *Chaoborus* are not known. Since *Chaoborus* individuals do not suffer from senescence until after their emergence, however, and since there is no evidence of food shortages sufficient to cause starvation, it seems safe to assume that fish predation accounts for essentially all mortality of *Chaoborus* larvae prior to pupation. The very large variation in *Chaoborus* production must therefore be due partly to variations in fish predation. Such variations are very likely related to the size structure of the *Chaoborus* population and to the presence or absence of a deoxygenated refuge zone for larger *Chaoborus* in deep water.

Another probable source of variability in *Chaoborus* production which may be even larger is the unpredictable reproductive success of adults at the time of emergence and the uneven dispersion of eggs resulting from the high mobility of the adults after emergence.

The failure of food abundance to account for a large percentage of the variance in *Chaoborus* production does not prove that food abundance is entirely unimportant in its effect on *Chaoborus* production in Lake Lanao. The data do suggest, however, that other factors, and probably mortality mechanisms, have such a variable effect on the population through time as to overwhelm the influence of food abundance as a source of variation.

Chapter 15

Community Trends in Growth and Reproduction

Figure 15–1 and Table 15–1 summarize the variation through time of energy flow (as net production) at the primary producer level, at the herbivore level, and at the primary carnivore level and the relationship between these variables. Figure 15–1 shows that the gap between the phytoplankton and herbivores is consistently greater than the gap between the herbivores and primary carnivores, indicating that energy transfer between the first and second trophic levels is consistently more inefficient than between the second and third trophic levels. The figure also suggests some relationship between energy flow at the first and second trophic levels and this is confirmed in Table 15–1. Although the relationship is far from perfect, it is clear that herbivores are influenced by primary production even though they ingest only a minor portion of the phytoplankton net production.

There is no significant relationship between the production of primary carnivores and the production of herbivores, which confirms the conclusion of the analysis of primary carnivore energy flow in the previous section. Further consideration will be given to overall control pathways in the concluding discussion.

The production data for each of the zooplankton herbivore categories can be used in an analysis of general community trends in growth and reproduction. The general working hypothesis for the analysis is that the community is structured according to definite rules which relate to the size, reproductive capacity, turnover rate, and other characteristics of individual species.

Three principal population characteristics are considered in turn and the results are then drawn together in a more general overview. These three characteristics are (1) absolute production (μg/liter/day). (2) turnover, or relative production (percentage/day), and (3) birth rate.

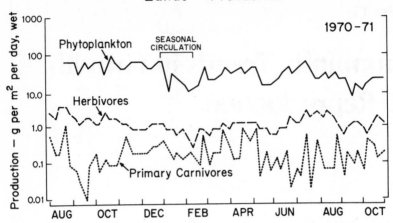

Figure 15–1. Comparison at weekly intervals of total energy flow at three trophic levels in Lake Lanao over the 65-week study period.

Trends in Absolute Production

The specific working hypothesis here is that the production of a population should be related in some way to the average size of the organisms which comprise it. The most likely relationship between body size and production a priori would seem to be one in which maximum population production is observed in organisms of intermediate size, and that organisms either smaller or larger than this intermediate size would experience increasing amounts of ecological stress and thus reduced average production.

To test the working hypothesis, the average production figures for all 12 species of zooplankton in the lake were plotted against the weighted average biomass per individual for each species. Because of the enormous range of values, both variables were log transformed prior to plotting. The plot re-

Table 15–1. Correlation Matrix for Energy Flow at Weekly Intervals in Three Trophic Levels over a 65-Week Period[a]

		Production		
		Phytoplankton	Herbivore	Primary carnivore
Production	Phytoplankton		0.47[b]	−0.05
	Herbivore			0.04
	Primary carnivore			

[a] Variables were log transformed.
[b] $P < 0.01$.

veals a remarkably uniform increase in production with weight (Fig. 15–2) which matches the similar trend in total mortality documented in Chapter 13. The log–log relationship is highly significant ($P < 0.01$) and explains 53% of the variance in production:

$$\log Y = -0.24 + 0.99 \log X$$

where Y is mean annual absolute production (μg/liter/day, wet) and X is wet weight (μg/individual).

Contrary to the original working hypothesis, Figure 15–2 provides no real evidence for a peak followed by a decline in production with increasing size. Furthermore, an examination of the details of production data (Table 9–1) confirms that if the major developmental classes of copepods are considered in relation to each other, there is no general decrease of production with size among the largest developmental stages. Thus it would appear that the trend relating production and size of individuals is most accurately described as monotonically increasing rather than peaked in the middle.

Figure 15–2 suggests a sort of paradox in which increasing size of individuals in a species leads to increasing production, and yet some factor appears to set a rather severe upper limit on the size of zooplankton herbivores. This

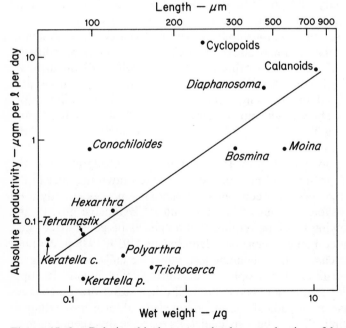

Figure 15–2. Relationship between absolute production of herbivore populations and the average size of individuals in the populations. The points and regression line are placed on the basis of wet weight per individual. The scale for body length is only approximate because of the imperfect relationship between length and body weight.

paradox can be discussed more meaningfully after the remaining portion of the analysis has been completed.

Figure 15–2 also provides a good opportunity for examination of the distribution of productivities in the community. Although there are insufficient numbers of species for construction of a proper frequency distribution, it is obvious that the frequency distribution on a linear scale would be skewed to the right, and that the total production of the community is dominated by a couple of species.

Figure 15–2 also illustrates a taxonomic trend in production. The two most productive species are copepods. Following these are the three cladoceran species, and last of all are the rotifer species. It is remarkable that there is no overlap between these groups.

Underlying the taxonomic trend is a life history trend. The organisms are arranged top to bottom on the productivity scale in a pattern which very closely reflects the ratio of biomass allocation to an individual before and after hatching. In the cyclopoids, for example, an average egg is 1.8% of the biomass of the average adult. Thus 1.8% of the biomass is accumulated prior to hatching, and the balance after hatching. At the other end of the spectrum are the rotifers. *Keratella cochlearis,* for example, produces an egg which is 66% of the average adult size. Thus the accumulation of biomass is 66% preinduction and only 34% postinduction. The cladocerans fall between the copepods and the rotifers in this respect.

Figure 15–3 shows the relationship between mean absolute production of species and the relative importance of preinduction and postinduction growth in the life history pattern (three rare rotifers with indistinguishable eggs are omitted). It is very clear from the figure that the relative importance of preinduction growth increases in direct relationship to the overall production of the species. The relationship depicted in Figure 15–3 is highly significant ($P < 0.001$; $\log Y = 1.77 - 1.62 \log X$) and accounts for 77% of the variance in production.

The full implications of Figure 15–3 cannot be considered until the conclusions of the present and previous sections are drawn together in the final discussion. One very clear ecological implication emerges directly from Figure 15–3, however, without additional context. Extensive growth and development following hatching require that a species be generalized, i.e., that it have extended niche dimensions. Hutchinson (1959, 1978 p. 216) has noted the extreme case of this phenomenon, which he calls metaphoetesis, in which the organism changes trophic levels during development. In copepods the trophic level may not change but still other rather drastic changes must occur because of the radical attenuation in size and form. For example, even though a nauplius 2 may be a specialist, its extreme divergence in morphology and size from a copepodid 1V of the same species ensures corresponding differences in resource requirements and mortality mechanisms, both in qualitative and quantitative terms. It has even been demonstrated

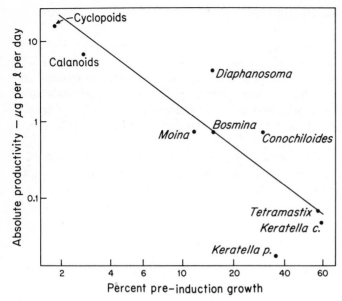

Figure 15–3. Relationship between absolute production of herbivore populations and the percentage of preinduction growth in the populations (i.e., the ratio of egg weight to adult weight).

that the size-metabolism relationship changes in a major way between nau-plius and copepodid (Epp and Lewis 1979a). In contrast, the dimensions of the niche space for a rotifer species would not be expected to change radi-cally through development because of the small amount of postinduction growth and the consequent stability in size and morphology. This effect is compounded by the interaction of environmental variation with duration of development to maturity. A large amount of postinduction growth implies a longer time span for development to age of first reproduction, which in turn implies exposure to a much greater range of environmental conditions prior to the age of first reproduction. Size change during development is of very great importance to the overall ecological requirements and vulnerabilities of zooplankton species, but has not often been appreciated sufficiently in the ecological analysis of zooplankton (however, see Neill, 1975 and Lynch, 1977).

   If greater amounts of postinduction growth do imply a broader niche, Fig-ure 15–3 in effect shows that successful generalists account for a great deal more production than successful specialists. A corollary, also supported by Figure 15–3, is that the transition from successful but unproductive special-ist to successful and highly productive generalist is a relatively smooth one which in part governs community structure.

Trends in Relative Production

In this analysis the major developmental stages of copepods are treated separately rather than being combined to give a mean value for the species. The reason for this is that any trends in the ratio of production to biomass are likely to derive from metabolic laws which apply within species as well as between species. The conclusions are nevertheless qualitatively the same if the developmental stages of copepods are lumped together rather than separated.

The specific working hypothesis in this case is that the ratio of production to biomass will bear an inverse relationship to the size of organisms. Figure 15–4 shows the relationship of P/B to size. The apparent trend downward in P/B ratios with increasing body length is statistically significant ($P < 0.01$) and thus supports the hypothesis. The relationship is:

$$Y = 27.6 - 0.023X$$

where Y equals the relative production (percentage/day) and X is the length of an average individual in microns. The relationship accounts for 44% of the

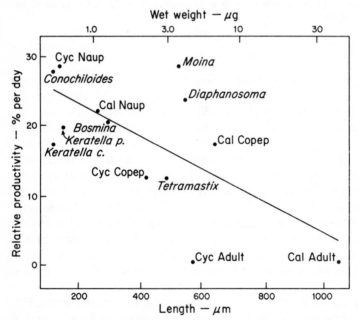

Figure 15–4. Relationship between relative production (i.e., turnover of biomass) of herbivore populations and the mean size of individuals in the populations. The placement of the points and the regression line are based on length, and the wet weight axis is only approximate because of the imperfect relationship between length and weight.

variance in production. The comparable relationship for wet weight is:

$$Y = 21.1 - 0.64X$$

where X is wet weight of an average individual in micrograms. In this case $P = 0.02$ and 35% of the variance in production is accounted for.

In the simplest case, the trend in P/B ratio with size would be explained exclusively on the basis of the metabolic laws which relate size to metabolic rate. The nature of the size–metabolism relationship among small metazoans is:

$$M = kB^b$$

where M is a measure of metabolic rate per individual, B is the biomass of the individual, and k and b are constants representing the relationship of metabolism to size.

The literature on poikilotherms indicates that the value of b under standard laboratory conditions using oxygen consumption as an index of metabolic rate will average very near 0.75 and will almost never fall outside the range 0.55 to 0.90 (Zeuthen, 1953; Hemmingsen, 1960). Epp and Lewis (1979b) have shown that this general rule is specifically applicable to tropical copepod populations. The equation, however, deals with metabolic rate per individual, whereas Figure 15–4 deals with metabolic rate (growth) per unit biomass. In order to make comparisons, we must put the general metabolism equation on a relative basis with respect to weight. To accomplish this, we simply divide both sides of the equation by weight per individual, which gives:

$$\frac{M}{B} = kB^{(b-1)}$$

where symbols are as given above. Now M/B is relative metabolism, or QO_2 if metabolism is measured by oxygen consumption. The term M/B will have dimensions metabolism/weight/time. The logarithm of both sides yields the equation:

$$\log (M/B) = \log k + (b - 1) \log B.$$

The empirical data in Figure 15–4 can be put into a form which is exactly the same form as the equation given above. When this is done, the empirical value of b in the Lanao plankton community can be compared with b as given by the general metabolic law. A line is fitted by least squares to the data in Figure 15–4 after log transformation of both variables. The result is:

$$\log (P/B) = 1.16 - 0.30 \log B.$$

Since P/B is a measure of metabolism relative to weight with dimensions metabolism/weight/time, then the constant 1.16 in this equation is comparable to the constant $\log k$ in the previous equation and the constant -0.30 is comparable to $b - 1$. Thus for the empirical relationship, $b - 1 = -0.30$, so $b = 0.70$ (standard error, ± 0.15). This empirically determined value of b is extremely close to the value given by the general metabolic law (0.75).

Apparently the ratio P/B in natural zooplankton communities over extended periods has never been examined in light of the metabolic law relating size to metabolism. The metabolic law and the M/B relationship derived above from it are based on measurements of gross metabolism (oxygen consumption), whereas the P/B ratio is based on the amount of net metabolism diverted to growth. It is not a foregone conclusion that the two different metabolism measures will show the same relationship to B. Since the b value derived from P/B analysis on Lake Lanao zooplankton does in fact agree remarkably well with the general b value for gross metabolism, however, some ecologically significant conclusions about the ratio of gross metabolism to growth are possible.

The analysis indicates no significant trend in the ratio of metabolism to growth, and thus suggests that the efficiency of herbivory does not change significantly with herbivore size. Efficiency in this context refers to the composite proportion of all metabolism diverted to growth under actual field conditions for entire populations. This does not mean that trends in amount of food harvested per unit body weight, assimilation efficiency, efficiency of movement, and other similar variables will not show trends with size, but rather that the composite effect of all trends is such that gross metabolic trend translates without significant change to a growth trend under field conditions. Studies of specific critical functions in copepods such as filtration and ingestion rates in fact seem to show varied results, with the functions in some instances regressing with size at about the same rate as general metabolism (e.g., Paffenhöffer, 1971; Paffenhöffer and Harris, 1976), but in other cases at a rate that would appear to be very different (e.g., Allan et al., 1977). The Lanao data suggest that, despite possible variations in specific functions, there is an overriding general tendency for growth of herbivores under field conditions to follow the general metabolic laws derived from laboratory studies outside the ecological context, which in turn implies no trend in efficiency of herbivory with size.

Examination of Figure 15–4 reveals that there is considerable variation around the trend line relating P/B to size. For example, the P/B for adult copepods is well below the line, while the two largest cladocerans fall well above the line. Individual species and developmental stages thus depart to various degrees from the predicted position based on size, even though a statistically significant trend exists. Consideration of community trends obviously does not entirely obviate the necessity for consideration of peculiarities of individual species and stages. The position of the adult copepods is especially significant and will be considered further in the next chapter.

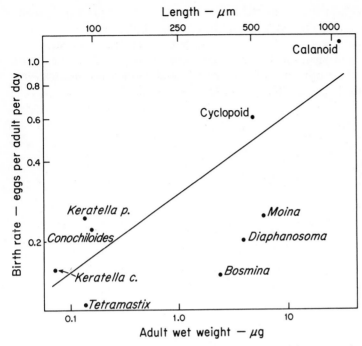

Figure 15-5. Relationship between birth rate of herbivore species in Lake Lanao and the size of breeding adults in the populations. Placement of the points and the regression line is based on wet weight and the length axis is only approximate because of the imperfect relationship between length and weight.

Trends in Birth Rate

After log transformation to reduce the scatter of points, the crude birth rate (births/adult/day) and the mean weight of adult organisms in the populations prove to be related. The relationship is illustrated in Figure 15–5 (log $Y = -0.58 + 0.23$ log X). The relationship is statistically significant ($P = 0.02$) and accounts for 35% of the variance in birth rates.

The results in this case are counterintuitive. It would appear that the birth rate of small organisms should be higher than the birth rate of larger organisms since the specific growth and mortality rates of smaller organisms are larger. Because of the radical increase in proportion of postinduction growth with size, however (Fig. 15–3), larger organisms are forced to sustain a higher birth rate to compensate for the much longer period over which mortality extends prior to the production of the first eggs. Thus while the mortality risk of a smaller species is greater for a fixed unit of time, the cumulative mortality between hatching and the age of first reproduction is much greater in larger organisms and must be offset by a higher birth rate.

In addition to the significant size trend in birth rates, there is also a taxonomic clustering of birth rates. The copepods show by far the highest birth rates, whereas the birth rates of cladocerans and rotifers are more evenly matched despite their considerable difference in size.

Chapter 16

Conclusion and Synthesis

Energy Flow and Biomass

Figure 16–1 shows the biomass and production pyramids for the Lake Lanao plankton system plus the transfer efficiencies between trophic levels. Transfer efficiency between the solar source and the phytoplankton community is relatively high in view of the low nutrient base in Lake Lanao. The summary of IBP data given by Brylinsky and Mann (1973) shows an average value of 0.4% for transfer of photosynthetically available solar energy (PAR, ca. 350–700 nm) to gross production. In terms of net production, which is of more interest here, this would be 0.2 to 0.3% based on PAR or about 0.1 to 0.15 based on total solar radiation. The percentage reported in Figure 9–1, which is based on total radiation, is obviously quite high. Efficient nutrient cycling by the mechanisms described in Chapter 3 is in large part responsible for minimization of severe nutrient depletion and consequent maintenance of high transfer efficiency at the first trophic level (Lewis, 1974).

Transfer efficiency between the phytoplankton and herbivore categories would appear to be low considering the richness of the phytoplankton food source and the sustained high temperatures. Actually, transfer efficiencies between 4 and 10% are not unusual in freshwater systems, however (e.g., Wright, 1958; Gulati, 1975; Rey and Capblancq, 1975; Makarewicz, 1975; Coveney et al., 1977). In Lake Lanao, the proximate explanation for low transfer efficiency is the low standing stock of herbivores (Fig. 16–1). The herbivore populations are sufficiently small that they cannot consume more than a small proportion of the total primary production in Lanao, as indicated in Chapter 14. Similar failure of herbivores to ingest most of the pri-

Lake Lanao

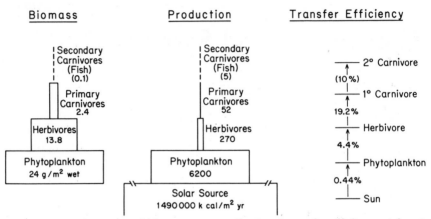

Figure 16–1. Summary of biomass, energy flow, and transfer efficiency at four trophic levels in Lake Lanao.

mary production has been documented recently for a variety of lakes (Rigler et al., 1974; Jassby and Goldman, 1974; Kalff et al., 1975; Coveney, et al., 1977).

Obviously it is very significant to the overall dynamics of the plankton community that the phytoplankton–herbivore linkage has such a low efficiency. The low efficiency is consistent with the theory of Gliwicz (1969) that eutrophic lakes characteristically show low transfer efficiencies between primary producer and herbivore levels, but low efficiencies have also been documented in oligotrophic lakes (Makarewicz, 1975). The analysis of Lake Lanao data has shown that both resource supply (food) and mortality (predation) affect the herbivores. The food resource appears not to be taxed by herbivory, as indicated by the small role that grazing plays in determining phytoplankton succession and the low proportion of total phytoplankton biomass which is ingested. Although the herbivores respond to changes in food quality brought about by physical–chemical factors, they give little evidence of exhausting even the preferred foods. The herbivore level appears to be heavily suppressed by predation, which primarily accounts for its failure to crop a larger proportion of the phytoplankton.

The transfer between herbivores and primary carnivores (*Chaoborus*) is extremely efficient (Fig. 16–1), and the analysis of predation has shown how the high efficiency is maintained. *Chaoborus* has a high ratio of production to biomass and a correspondingly high rate of food intake per unit biomass. *Chaoborus* predation is thus sufficient to account for about 95% of total herbivore losses. *Chaoborus* concentrates on the herbivore categories where production is highest, thus diverting maximum possible herbivore production into its own growth. Although the failure of *Chaoborus* to harvest many of the small rotifer species at first appears puzzling, it is clear from energetic considerations that this food source is insignificant. Since absolute production among herbivore species is correlated in Lake Lanao with mean body size, it is obviously advantageous for primary carnivores to specialize in herbivores of moderate to large size, which is precisely the strategy followed by *Chaoborus*. *Chaoborus* seems to be somewhat imprudent (sensu Slobodkin 1961) in its use of herbivores, as it prevents maximal harvest of phytoplankton foods by suppression of herbivore populations. On the other hand, *Chaoborus* shows some of the more subtle earmarks of a prudent predator by concentrating on larger zooplankton, thus allowing the smallest developmental stages with their high relative growth rates to harvest the plant foods (Slobodkin 1974).

The key contrast between transfer efficiency from trophic level one to two as compared with efficiency from trophic level two to three lies in the amount of food ingested. Herbivory is less efficient than primary carnivory mainly because the herbivores are suppressed to the extent that they cannot ingest a major portion of their food resource.

Magnitude of the transfer between primary carnivores (*Chaoborus*) and secondary carnivores (fish) is known on the basis of *Chaoborus* mortality, but no transfer coefficient can be calculated because fish growth was not studied. Mortality in the primary carnivore suggests that predation by secondary carnivores definitely affects the population dynamics of the primary carnivore but probably does not directly suppress production at the primary carnivore level. This impression is reinforced by the documented ability of *Chaoborus* to harvest essentially all of the herbivore production. Fish thus appear to have a prudent predator relationship with *Chaoborus* (Slobodkin 1961).

Since a large production of *Chaoborus* biomass is diverted to fish, the resultant fish yield can be approximated given some assumptions about the metabolic efficiencies of the fish. Assuming a 10% transfer efficiency to the secondary carnivores, *Chaoborus* production alone would support an annual fish production of 103 kg/ha (wet weight). This is presumably augmented by benthic fish production nourished by energy passing through the detrital food chain. According to Melack's (1976) relationship relating observed fish yields to gross primary production in tropical lakes, Lake Lanao should yield about 60 kg/ha/year to a fishery. Since production must always exceed yield to man, the computed production appears to agree rather well with Melack's relationship.

Both biomass and energy flow for all major trophic levels show low relative variability through time (Table 16–1). This is accounted for by the absence of harsh physical conditions comparable to those which would occur in the temperate zone.

The relative variability in energy supplied by the solar source is consider-

Table 16–1.   Relative Variability on a Weekly Time Scale of
Biomass and Energy Flow in the Plankton Trophic Levels of
Lake Lanao

Trophic level	Relative variability— Coefficient of variation (%)	
	Biomass	Production
Solar source	—	15
Phytoplankton	61	51
Herbivores	48	57
Primary carnivores	122	104

ably lower than relative variability in energy flow at the phytoplankton level.
Some variance at the first trophic level is induced by changes in incident
light but this is greatly augmented by changes in nutrient availability and by
changes in depth of mixing and transparency that modify the distribution of
incident light to the primary producers (Lewis, 1978b). These resource supply factors account for most of the variance in primary production (Lewis,
1974).

Variability in energy flow at the phytoplankton level is very similar to variability at the herbivore level. Energy flow through the carnivore component,
however, is considerably more variable. Variation in the biomass of given
trophic levels is very similar in magnitude to variation in energy flow in the
same trophic levels.

Community Structure and Organization

The structure and organization of the Lake Lanao plankton system are quite
simple in outline. The phytoplankton assemblage of Lake Lanao is composed of unicellular, colonial, or coenobial organisms (usually 1–25 cells
each). Approximately 90% of the individuals have a wet weight between
0.00002 and 0.002 μg. At the herbivore level, approximately 90% of individuals fall within the range 0.1 to 20 μg (wet). For the primary carnivores,
the corresponding weight range is 3.5–3500 μg per individual. Thus individuals of each trophic level are nourished primarily by particles of much
smaller size from the trophic level below.

The primary producer level is by far the most diverse, as it contains some
70 species from five major taxa. The herbivores are second with 12 species in
three major taxa and some additional diversification provided by the extremely great morphological changes in copepods during development. The
primary carnivores are least diverse of all, as they are limited almost entirely
to one species of the genus *Chaoborus*.

For a more mechanistic view of community structure, it is instructive to consider factors which are likely to determine the size boundaries of each of the major trophic levels. For phytoplankton, nutrient transport phenomena and sinking strongly influence the size and shape of the biomass unit. These factors appear to be largely responsible for conservation of the surface to volume ratio within definite limits even for very different species (Lewis, 1976) and for the placement of species in the successional sequence (Chapter 3; Lewis, 1978b). Vulnerability to grazing is not strongly related to volume of the biomass units, but is probably related to their dimensions and shape as shown in Chapter 14. The effect of grazing on the size structure and size limits of the phytoplankton is reduced, however, by the inability of herbivores to crop a large proportion of phytoplankton biomass. Both the upper and lower size limits of phytoplankton are thus to a large extent determined by physical and chemical requirements and only secondarily by grazing.

The lower size range of herbivore species includes small rotifers and the early naupliar stages of *Thermocyclops*. Given the requirement for considerable morphological complexity in connection with feeding and reproduction, these organisms probably approach an absolute lower size limit for metazoans. Protozoans, although not considered here, comprise a measurable percentage of total heterotroph biomass (Lewis, 1974) and, although they are quantitatively even less significant as grazers than the rotifers, do engage in herbivory.

Factors enforcing an upper size limit for zooplankton herbivores are enigmatic, as it would appear that large herbivores are highly successful, yet the size distribution is abruptly truncated at a size of about 1 mm. The size structure of the community suggests that some kind of transition occurs in the feasibility of herbivory when herbivores reach a body weight of approximately 30 μg per individual or a length of approximately 1 mm.

There is general evidence to support the hypothesis that a structural transition occurs very widely in freshwater communities as a result of intensive vertebrate predation on organisms 1 mm or greater in size (see Brooks, 1968; Hall et al., 1976). This contention can in fact be supported empirically for Lake Lanao. Predation losses of Lanao herbivores, all of which are < 1 mm in size, are due almost entirely to *Chaoborus*, an invertebrate predator. In contrast, *Chaoborus* itself, which begins life at a size of almost exactly 1 mm, experiences heavy losses due to fish predation. The herbivores apparently find it ecologically advantageous to reach a maximum size of less than 1 mm and thus avoid vertebrate predators, which then concentrate on the primary carnivores. Reinforcing the upper size limit on herbivores is the decrease in P/B ratio with size in the herbivore class according to the metabolic law, which renders larger herbivores less able to withstand a given predation pressure.

The lower size limit of primary carnivores is obviously determined in large part by minimum size necessary to permit capture and manipulation of nutritionally suitable prey. It would appear that the smallest *Chaoborus* larvae in

Lake Lanao are just large enough to satisfy this criterion. The youngest *Chaoborus* have empty crops much more often than older ones (Lewis, 1977a), suggesting that the youngest animals are not able to feed ad libitum because of their small size. The upper size limit for primary carnivores is determined by a balance between the disadvantages of extended larval development in the face of fish predation and the disadvantages in terms of fecundity and reduced viability of an unduly small adult stage. A careful study of the mortality and reproductive success of emerging adults would probably show that the size of *Chaoborus* is adjusted to the severe physical challenges associated with emergence and egg-laying on a water surface exposed to the wind.

Rules Governing Structure of the Herbivore Component

A more detailed causal picture of community structure among the herbivores is possible through synthesis of the demographic trends that have been documented in herbivores. Figure 16–2 summarizes these trends. Both size and demographic factors are plotted on linear scales, and the limits in each case are set by the maximum and minimum values observed in the community. The trend lines, which are based on the actual analyses of mortality and growth, sometimes do not pass exactly from one corner of the graph to another because there is a scatter of points around the lines. The trend lines for mortality and production in Figure 16–2 are essentially the same, as would be expected in view of the fact that mortality and production balance for any species over an extended interval of time.

Given the detailed analysis of mortality and growth control mechanisms plus trends illustrated in Figure 16–2, it is possible to specify some of the ecological and evolutionary rules governing plankton community structure and organization.

Rule 1

The minimum feasible metabolic rate of herbivores generally decreases with increasing size, while the predation pressure on herbivores increases with increasing size.

The analysis of P/B ratios has shown that the overall efficiency of herbivory is unaffected by body size. Metabolic laws relating metabolism to size dictate that if efficiency is constant, both gross and net metabolism will decrease with increasing size. This implies that the minimum food-gathering requirements per unit body weight will also decrease with size. A reasonable corollary of this is that resource requirements of herbivores decrease in

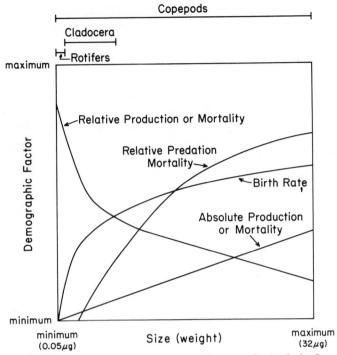

Figure 16–2. Summary of demographic trends in the Lake Lanao plankton community. Each of the demographic factors in the figure is scaled according to the minimum and maximum values of that particular factor among the species in the Lake Lanao plankton community.

stringency with increasing body size. Resource requirements in this context may be interpreted jointly in terms of quantity and quality. As an example, let us suppose that a large herbivore and a small one both rely on four out of 50 phytoplankton species for their principal growth and maintenance (Chapter 14). The P/B analysis shows that the rate of metabolism and growth must be faster in the small herbivore, but that the composite efficiency of harvesting and processing foods is the same. If the critical phytoplankton species decline in abundance at a certain time of year, the critical density to support the small herbivore will be reached before the critical density to support the large herbivore. The large herbivore will thus be resource limited to a much smaller degree than the small herbivore. The critical point here is that the small herbivore, for ecological or physiological reasons, does not have the option of facultative reduction in metabolism and is thus locked to a higher minimum resource requirement. This explains why small herbivores, which suffer very limited predation loss, do not dominate the biomass or production of the community. The increasingly restrictive requirements with decreasing size are evidently sufficient to offset reduced predation on small herbivores, although the balance might easily shift in response to changes in either the mean predation pressure or the mean resource supply.

Rule 2

The ratio of production to mortality is strikingly unfavorable specifically for the adult stage of large, sexually reproducing herbivores, indicating that the cost of sexual reproduction is so high that it can be borne only by large herbivores.

The nature of the balance between production and mortality is summarized in Table 16–2 for copepods, which span a very broad size range. Table 16–2 shows that the ratio of production to mortality differs greatly between age (size) groups of both copepod species. For a species to survive, the relative mortality rate over any extended period of time must match the relative production for the sum of all the developmental stages. Thus if the relative mortality of adults is to be higher than their relative production, this inequality must be balanced by the reverse relationship in earlier developmental stages. This is in fact the case for both copepod species.

Table 16–2 also shows quite clearly why adult copepods do not produce large eggs. A copepod which produced eggs weighing 50 to 75% of its own body weight would quickly become extinct because of the extremely unfavorable ratio of mortality to production in the adults. Relative mortality and production need not be balanced within developmental stages, hence by producing small eggs, copepods can take advantage of the more favorable ratio of production to mortality in earlier developmental stages and thus accumulate the energy required to sustain the adult stage.

The pattern of production/mortality ratios through development in the two copepods is different, as might be expected from the analysis of their life

Table 16–2. Ratio of Relative Production to Relative Mortality through Development in Copepods

	Cyclopoids			Calanoids		
	Nauplius	Copepodid	Adult	Nauplius	Copepodid	Adult
Relative production (%/day)	29.0	12.7	0.9	22.3	17.7	1.3
Relative mortality (%/day)	1.6	15.6	5.0	18.5	9.6	8.0
Ratio, production/ mortality	18.1	0.8	0.2	1.2	1.8	0.2

histories. For the cyclopoids, relative production greatly exceeds relative mortality in the naupliar stage, but mortality and production are almost evenly matched in the subadult copepodid stages. For the calanoids, production and mortality are very evenly matched in the naupliar stages and the subadult copepodid stages show a distinct excess of production over mortality.

It is clear from the table that the adult copepods in particular experience a markedly unfavorable balance of mortality and production. The decline in relative production between the subadult copepodid stages and the adult stage for both species is so precipitous as to indicate that the low relative production of adults is in large part connected specifically with sexuality rather than with size alone. The very low relative production of adults may derive from the "twofold cost" (Maynard-Smith, 1978) of sexual reproduction in contrast to parthenogenesis due to the maintenance of two sexes. Maintenance of sexuality obviously requires a considerable genetic advantage to offset costs (Williams, 1975, Maynard-Smith, 1978).

The high cost of sexuality provides a needed explanation for the failure of natural selection to reduce the size of the adult copepods. If the very unfavorable productivity/mortality ratios of adults indeed derive from sexual reproduction then it is probable that the energy costs of sexual reproduction in smaller herbivores would be completely prohibitive due to their much higher relative metabolic rate, which if compounded by a drastic decrease in efficiency resulting from sexuality would result in an enormous resource requirement per unit body weight. This could be offset by a drastic decrease in the ratio of egg to adult size, as seen in the copepods, but the resulting offspring would be extremely small. In *Keratella,* for example, adoption of an egg:adult ratio of 2% as seen in copepods would result in young individuals weighing only 0.001 μg. Such a size reduction would obviously invalidate feeding adapations based on the phytoplankton size spectrum and would presumably imply extremely high metabolic rates in young individuals.

Rule 3

The unfavorable production/mortality ratio of adults in larger sexually reproducing herbivores, which requires them to produce small eggs, also requires them to sustain high birth rates.

One striking trend in Figure 16–2 is the increase in birth rate with size in the Lake Lanao herbivore species. The very large difference in size between the egg and the adult in large herbivores requires an extended developmental period. Even when mortality rates are low, lengthy development extends the mortality risk over a much longer period and thus requires a higher birth rate to supply the necessary number of individuals reaching age of first reproduction.

Rule 4

Trends toward increasing difference in size between eggs and adults and higher birth rates in larger herbivores require a corresponding increase in absolute production of herbivores with size.

Sexually reproducing organisms must maintain some minimal density in order to survive. The foregoing arguments show that for the largest herbivores a large number of eggs and a large number of developing organisms are required to maintain a small number of adults because of the extended duration of development. This means that in order to exist at all, large sexually reproducing herbivores must have a compensatory high absolute production. The extent of compensation depends on the intensity of mortality during development, which is in turn dependent on the efficiency of the primary carnivores. Since the primary carnivore *Chaoborus* is abundant in Lake Lanao, both of the copepod species must sustain high absolute production in order to generate an adequate number of breeding adults.

In summary, it is clear that metabolic factors, an extended developmental period, and sexuality could together counterbalance predation, the primary disadvantage of increasing body size in herbivores. This explains how a range of morphologies, sizes, and life history patterns can lead to evolutionarily stable strategies in the zooplankton.

Control Pathways in the Ecosystem

A composite overview of the mechanisms at work in the Lake Lanao plankton system can be conceived in terms of control pathways operating between major components of the system. Control may apply either to quantity (mainly energy flow and population dynamics) or to quality (mainly community structure and organization).

Figure 16–3 summarizes the control pathways in the Lake Lanao plankton system. The figure represents both quantity control and quality control, which differ considerably from each other in some cases, and also distinguishes between major and minor control pathways.

The intimate dependence of light availability, nutrient availability, and turbulence upon seasonal and nonseasonal weather patterns is represented in Figure 16–3 by the flow of control from weather to light, nutrients, and turbulence. This control is principally quantitative. Strong reciprocal control mechanisms then link the phytoplankton with their resource supply. This control is both quantitative and qualitative, since light and nutrient levels determine biomass and energy flow as well as succession and composition of the phytoplankton. Turbulence, which is related to mortality through sink-

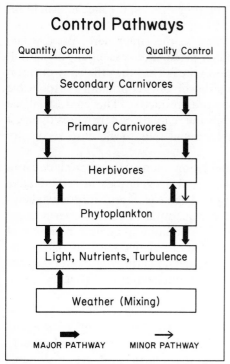

Figure 16–3. Control pathways for the Lake Lanao plankton community as deduced from the analysis in the text.

ing, has similar effects but is not reciprocally affected by the phytoplankton as light and nutrients are.

A considerable amount of evidence presented here demonstrates the control of growth and reproduction in herbivores by the abundance and quality of phytoplankton foods. Changes in phytoplankton abundance resulting principally from changes in abiotic factors are accompanied by shifts in phytoplankton species composition which are in turn critical to the control of growth and reproduction in herbivores because herbivores differ in their food requirements. The control in this case is not markedly reciprocal. Herbivores ingest only a small proportion of the phytoplankton crop and thus do not constitute a quantitatively important attrition control mechanism for phytoplankton. Since herbivore feeding is specific, however, herbivores do exercise a measurable amount of quality control on phytoplankton. Grazing is nevertheless consideraby less important than light, nutrients, and turbulence in determining phytoplankton community composition, hence even the qualitative control pathway flowing from herbivores to phytoplankton is shown as a minor pathway.

As indicated in the analysis of predation mortality, primary carnivores exercise important quality control on herbivores through the mechanism of selective predation and the trend toward higher predation rates on larger organisms. The analysis has also shown that in the larger herbivores, which account for the vast majority of secondary production, essentially all mortality can be accounted for by primary carnivores. Thus the efficiency of carnivores in removing the standing crop of herbivores is primarily responsible for the quantitative limitations on herbivore production. This control linkage is extremely important, as it accounts for the inability of herbivores to ingest more of the phytoplankton standing crop and thus to increase secondary production.

The analysis has also shown that population fluctuations in primary carnivores are more influenced by predation and irregularities in reproductive success than by food supply mechanisms. High *Chaoborus* mortality shows that secondary carnivores exercise strong quantitative control over primary carnivores, as indicated in Figure 16–3. In addition, predation mortality of large primary carnivores is twice as high on a relative basis as that of the smallest primary carnivores, hence the secondary carnivores exercise a considerable influence over the size composition (quality) of primary carnivores.

The major control pathways are to a large degree convergent on the herbivore component of the Lake Lanao system. This component absorbs and integrates the selective changes in phytoplankton composition and primary carnivores, both of which are in turn under the control of factors not affected by herbivores to any great degree. Flow of control in the animal component of the community is downward from the topmost trophic levels, and this is matched by an upward flow of control from abiotic factors to herbivores.

Summary

The zooplankton community of Lake Lanao, Philippines, was studied over a 65-week period during which weekly samples were taken at multiple depths and multiple stations. All species and developmental stages were counted separately, and supporting data on physical–chemical properties of the lake and on the phytoplankton were also taken at weekly intervals.

The pelagic zooplankton community consists of one *Chaoborus* species, one calanoid and one cyclopoid species, four cladoceran species, and seven rotifer species. All of these are herbivores except *Chaoborus*. Annual mean zooplankton biomass is 16.1 g/m^2 (wet), and the comparable figure for phytoplankton biomass is 23.7 g/m^2. Zooplankton biomass declines during the seasonal mixing period, as does phytoplankton biomass.

Development rates for cyclopoids, calanoids, and *Chaoborus* were determined by cohort analysis supplemented by laboratory studies of egg development. Mean growth rates for cyclopoids range from about 40%/day (instantaneous rate) in the early naupliar stages to about 10%/day in the late copepodid stages. Cohorts differ somewhat in growth rate. Comparable figures for calanoid copepods are about 30%/day for early naupliar stages and about 15%/day for late copepodid stages. *Chaoborus* growth averages about 30%/day in the first instar and about 15, 10, and 5%/day in the succeeding three instars. Egg development times for all copepods, Cladocera, and rotifers are very near 1 day.

Secondary production was calculated for all species and stages based on the development rates and abundances. The cyclopoid copepod *Thermocyclops* dominates secondary production (46% of the total). Other groups and their percentages are: calanoids, 20%; all Cladocera, 15%; all rotifers, 2.8%; *Chaoborus*, 16%. Total herbivore production for the year averages 34.4 μg wet weight/liter/day, and total carnivore production averages 6.3 μg/liter/day. Total herbivore production is 4% of total net primary production.

Similarities in the temporal distribution of species and developmental stages were analyzed on the basis of four correlation matrices: the abundance and productivity matrices (N, P) and the matrices of rates of change of abundance and productivity $(\Delta N, \Delta P)$. The N matrix focuses on community structure, whereas the other three matrices focus on community dynamics. The ΔN matrix provides the best overall measure of ecological similarity in growth and mortality, whereas the P matrix is the best measure of similarity in timing of demands on algal resources. The ΔP matrix emphasizes similarity in growth control mechanisms. Measures that are sensitive to similarity in population dynamics show that the developmental stages of copepod species are tightly linked, but this linkage does not extend to community structure. In terms of community dynamics, the Cladocera have a strong tendency to group with each other. The rotifers are a remarkably coherent group both in terms of community dynamics and community structure.

A detailed analysis was made of mortality for all species and developmental stages. For the copepods and *Chaoborus,* which have extended life histories, mortality was determined with a week-by-week simulation of population growth over successive short time increments (0.1 day) using the known abundances and growth rates. Mortality was computed as the difference between simulated (no mortality) and observed population growth. A similar but simpler simulation was used for cladocerans and rotifers. Average mortality rates vary between 5 and 26%/day for herbivores except cyclopoid nauplii, which show exceptionally low mortality (1.6%/day). The mortality of *Chaoborus* ranges between 2 and 6%/day depending on instar. Since the feeding electivities of the *Chaoborus* population are known, it can be shown that essentially all of the mortality of copepods and Cladocera can be accounted for by *Chaoborus* predation. The mortality of five of the seven rotifer species cannot be accounted for by predation of any kind. An analysis of this unexplained mortality in rotifers shows that it is statistically related to the species composition of the phytoplankton community, suggesting that food supply is related to mortality. The community as a whole shows a significantly negative relationship between relative total mortality (percentage/day) and body size. There is a statistically significant opposite trend in relative mortality specifically attributable to predation, demonstrating a shift in mortality control mechanisms with size.

Mechanisms controlling growth and reproduction of all species and developmental stages were analyzed on the basis of production. Comparison of herbivore production with the quality and quantity of the phytoplankton food source explains a very large amount of the variance in herbivore production. All herbivore species and developmental stages show one or more statistically definable relationships to specific phytoplankton taxa at the class or division level. Multivariate comparison of the production of individual herbivore types with abundances of individual phytoplankton species shows that only a few phytoplankton species are important regulators of herbivore growth. Green algae in particular are generally unsuitable for herbivores. A statistical analysis of the relationship between herbivore production and phytoplankton size fractions demonstrates that herbivore production is not coherently related to the availability of particles of a certain volume range. There is a significant tendency, however, for copepod production to be related to the availability of food items which are elongate and have small cross-sectional measurements, and for the production of Cladocera to be related to the availability of phytoplankton foods which are shorter and have small cross-sectional measurements. The rotifers show no overall coherence with respect to particle shape. Whereas the production of herbivores is closely related to resource supply, variation in the production of the predator *Chaoborus* is essentially independent of food abundance and is instead very strongly under the control of mortality mechanisms and irregularities in the reproductive success of the adults.

Among herbivores there is a statistically significant trend toward higher absolute annual production with increasing body size and with increasing amount of postinduction growth. There is also a significant downward trend in the production to biomass ratio with increasing size, and the steepness of the trend is predicted almost exactly from general metabolic laws relating metabolism to size, suggesting that efficiency is essentially unaffected by size in these herbivores.

The community trends and control mechanisms form a coherent picture of community structure and energy flow. Control mechanisms are convergent on the herbivore trophic level, which is in turn composed of species that differ radically in size and life history. The duration of development, size and age at the time of first reproduction, ratio of production to biomass, and presence or absence of sexuality can be drawn together in an explanation of community structure and the simultaneous existence of several evolutionarily stable strategies in the community.

References

Allan, J. D. 1976. Life history patterns in zooplankton. *Amer. Natur.* **110**(971):165–180.

Allan, J. D., S. Richman, D. R. Heinle, and R. Huff. 1977. Grazing in juvenile stages of some estuarine calanoid copepods. *Mar. Biol.* **43**:317–331.

Berg, K. 1937. Contributions to the biology of *Corethra* Meigen (*Chaoborus* Lichtenstein). *Klg. Danske Videnskab. Selskab, Biol. Medd.* **13**:1–101.

Berman, M. S., and S. Richman. 1974. The feeding behavior of *Daphnia pulex* from Lake Winnebago, Wisconsin. *Limnol. Oceanogr.* **19**:105–109.

Bosselmann, S. 1975. Production of *Eudiaptomus graciloides* in Lake Esrom, 1970. *Arch. Hydrobiol.* **76**:43–64.

Bottrell, H. H., A. Duncan, C. M. Gliwicz, E. Grygierek, A. Herzig, A. Hilbricht-Ilkowska, H. Kurasawa, P. Larsson, and T. Weglenska. 1976. A review of some problems in zooplankton production studies. *Norwegian J. Zool.* **24**:419–456.

Brehm, V. 1933. Einige neue Diaptomiden. *Zool. Anz.* **103**:295–304.

Brooks, J. L. 1968. The effects of prey size selection by lake planktivores. *Syst. Zool.* **17**:272–291.

Brylinsky, M., and K. H. Mann. 1973. An analysis of factors governing productivity in lakes and reservoirs. *Limnol. Oceanogr.* **18**(1):1–14.

Burgis, M. J. 1970. The effect of temperature on the development time of eggs of *Thermocyclops* sp., a tropical cyclopoid copepod from Lake George, Uganda. *Limnol. Oceanogr.* **15**:742–747.

Burgis, M. J. 1971. The ecology and production of copepods, particularly *Thermocyclops hyalinus,* in the tropical Lake George, Uganda. *Freshwater Biol.* **1**:169–192.

Burgis, M. J. 1974. Revised estimates for the biomass and production of zooplankton in Lake George, Uganda. *Freshwater Biol.* **4**:535–541.

Burns, C. W. 1968. The relationship between body size of filter feeding cladocera and the maximum size of particles ingested. *Limnol. Oceanogr.* **13**:675–678.

Cassie, R. M. 1971. Sampling and subsampling: Theory and practice. In W. T. Edmondson (ed.), *A Manual on Methods for the Assessment of Secondary Productivity in Fresh Waters,* pp. 174–209. Blackwell, London.

Caswell, H. 1972. On instantaneous and finite birth rates. *Limnol. Oceanogr.* **17**:787–791.

Coker, R. E. 1943. *Mesocyclops edax* (S. A. Forbes), *M. leuckarti* (Claus) and related species in America. *J. Elisha Mitchell Soc.* **59**:181–200.

Comita, G. W. 1972. The seasonal zooplankton cycles, production and transformations of energy in Severson Lake, Minnesota. *Arch. Hydrobiol.* **70**:14–66.

Connell, J. H. 1978. Diversity in tropical rain forest and coral reefs. *Science* **199**: 1302–1310.

Conover, W. J. 1971. *Practical Nonparametric Statistics.* John Wiley, New York. 462 pp.

Cook, R. E. 1977. Raymond Lindeman and the trophic-dynamic concept in ecology. *Science* **198**:22–26.

Coveney, M. F., G. Cronberg, M. Enell, K. Larsson, and L. Olofsson. 1977. Phytoplankton, zooplankton and bacteria—standing crop and production relationships in a eutrophic lake. *Oikos* **29**:5–21.

Cressa, C. 1971. Distribucion vertical de la poblacion de *Chaoborus* sp. en el Embalse de Lagartijo (Estado Miranda) y su variacion estacional. Thesis de Grado, Universidad Central de Venezuela, Caracas. 48 pp.

Cummins, K. W., R. R. Costa, R. E. Rowe, G. A. Moshiri, R. M. Scanlon, and R. K. Zajdel. 1969. Ecological energetics of a natural population of the predaceous zooplankter *Leptodora kindtii* (Focke) (Crustacea; Cladocera). *Oikos* **20**:189–223.

Denman, R. L. 1975. Spectral analysis: A summary of the theory and techniques. Canada Dep. Environ. Tech. Report 539.

Doohan, M. 1973. Energetics of planktonic rotifers applied to populations in reservoirs. Ph.D. Thesis, University of London. 226 pp.

Eckstein, F. 1936. Beiträge zur Kenntnis exotischer Chaoborinae (Corethrinae auct) nebst Bemerkungen über einige einheimische Formen. *Arch. Hydrobiol. Suppl.* **14**:484–505.

Edmondson, W. T. 1965. Reproductive rate of planktonic rotifers as related to food and temperature in nature. *Ecol. Monogr.* **35**:61–111.

Edmondson, W. T. 1968. A graphical model for evaluating the use of egg ratio for measuring birth and death rates. *Oecologia* **1**:1–37.

Edmondson, W. T. 1971. *A Manual on Methods for the Assessment of Secondary Production in Freshwaters.* IBP Handbook No. 17. Blackwell, Oxford. 358 pp.

Edmondson, W. T. 1972. Instantaneous birth rates of zooplankton. *Limnol. Oceanogr.* **17**:792–795.

Eichhorn, R. 1957. Zur Populationsdynamik der calanoiden Copepoden in Titisee und Feldsee. *Arch. Hydrobiol. Suppl.* **24**:186–246.

Einsle, U. 1970. Études morphologiques sur des espèces de *Thermocyclops* (Crust. cop.) d'Afrique et d'Europe. *Cah. O.R.S.T.O.M. Sér. Hydrobiol.* **IV**:2:13–38.

Elster, H. J. 1936. Einige biologische Beobachtungen an *Heterocope borealis* Fisher (= *Weismanni imhof.*). *Int. Rev. Ges. Hydrobiol.* **33**:357–433.

Elster, H. J. 1954. Uber die Populationsdynamik von *Eudiaptomus gracilis* Sars und

Heterocope borealis Fischer im Bodensee-obersee. *Arch. Hydrobiol. Suppl.* **20:**546–614.

Elster, H. J. 1955. Zooplankton ein Beitrag zur Produktionsbiologie des Zooplanktons. *Verh. Int. Verein. Limnol.* **12:**404–411.

Epp, R. W. and W. M. Lewis, Jr. 1979a. The nature and ecological significance of metabolic changes during the life history of copepods. *Ecology* (in press).

Epp, R. W., and W. M. Lewis, Jr. 1979b. Metabolic responses to temperature change in a tropical freshwater copepod (*Mesocyclops brasilianus*) and their adaptive significance. *Oecologia* (in press).

Fairchild, G. W., R. S. Stemberger, L. C. Epskamp, and H. A. Debaugh. 1977. Environmental variables affecting small-scale distributions of five rotifer species in Lancaster Lake, Michigan. *Int. Rev. Ges. Hydrobiol.* **62:**511–521.

Frey, D. G. 1969. A limnological reconnaissance of Lake Lanao, Philippines. *Verh. Int. Verein. Limnol.* **17:**1090–1102.

Friedman, M. M., and J. R. Strickler. 1975. Two more receptors and feeding in calanoid copepods (Arthropoda: Crustacea). *Proc. Natl. Acad. Sci. U. S. A.* **72:** 4185–4188.

Fryer, G. 1957a. Freeliving freshwater Crustacea from Lake Nyasa and adjoining waters. Part 1. Copepoda. *Arch. Hydrobiol.* **53:**62–86.

Fryer, G. 1957b. The food of some freshwater cyclopoid copepods and its ecological significance. *J. An. Ecol.* **26:**263–268.

Galkovskaya, G. A. 1965. Planktonnye kolovratki i ikh rol v produktivnosti vodoemov. Dissert. Thesis. Biel. Gos. Univ. im. Lenina, Minsk, 1–19. (Cited in Winberg, 1971.)

Gerritsen, J. 1978. Instar-specific swimming patterns and predation of planktonic copepods. *Verh. Int. Ver. Limnol.* **20:**2531–2536.

Gerritsen, J., and J. R. Strickler. 1977. Encounter probabilities and community structure in zooplankton: A mathematical model. *J. Fish. Res. Bd. Can.* **34:**73–82.

Gilbert, J. J., and P. L. Starkweather. 1977. Feeding in the rotifer *Brachionus calyciflorus*. I. Regulatory mechanisms. *Oecologia* **48:**125–131.

Gliwicz, Z. M. 1969. Studies of the feeding of pelagic zooplankton in lakes with varying trophy. *Ekol. Polska Ser. A.* **17:**663–708.

Goulden, C. 1968. The systematics and evolution of the Moinidae. *Trans. Amer. Phil. Soc.* **58:**1–101.

Green, J. 1967. The distribution and variation in *Daphnia lumholtzi* (Crustacea: Cladocera) in relation to fish predation in Lake Albert, East Africa. *J. Zool. London* **151:**181–197.

Green, J. 1972. Latitudinal variation in associations of planktonic Rotifera. *J. Zool. London* **167:**31–39.

Green, J., S. A. Corbet, E. Watts, and O. B. Lan. 1976. Ecological studies on Indonesian lakes. Overturn and restratification of Ranau Lamongan. *J. Zool. London* **180:**315–354.

Gulati, R. D. 1975. A study on the role of herbivorous zooplankton community as primary consumers of phytoplankton in Dutch lakes. *Verh. Int. Ver. Limnol.* **19:**1202–1210.

Hall, D. J. 1964. An experimental approach to the dynamics of a natural population of *Daphnia galeata mendotae*. *Ecology* **45:**94–112.

Hall, D. G., S. T. Threlkeld, C. W. Burns, and P. H. Crowley. 1976. The size efficiency hypothesis and the size structure of zooplankton communities. *Annu. Rev. Ecol. Syst.* **7:**177–208.

Hargrave, B. T., and G. H. Geen. 1970. Effects of copepod grazing on two natural phytoplankton populations. *J. Fish. Res. Bd. Can.* **27:**1395–1403.

Hemmingsen, A. M. 1960. Energy metabolism as related to body size and respiratory surfaces, and its evolution. *Rep. Steno. Hosp.* **9:**1–110.

Hobbie, J. E., and P. Rublee. 1975. Bacterial production in an arctic pond. *Verh. Int. Verein Limnol.* **19:**466–471.

Hutchinson, G. E. 1959. Homage to Santa Rosalia; or, Why are there so many kinds of animals? *Amer. Natural.* **93:**145–59.

Hutchinson, G. E. 1967. *A Treatise on Limnology,* Vol. II. John Wiley, New York. 1155 pp.

Hutchinson, G. E. 1978. *An Introduction to Population Ecology.* Yale University Press, New Haven. 260 pp.

Hutchinson, G. E., and H. Loffler. 1956. The thermal classification of lakes. *Proc. Natl. Acad. Sci. U. S. A.* **42:**84–86.

Iltis, A., and P. Compere. 1974. Algues de la region du Lac Tchad. I. Caracteristiques generale du milieu. *Cah. O.R.S.T.O.M. Sér. Hydrobiol.* **8**(3/4):141–164.

Infante, A. 1978. Natural food of herbivorous zooplankton of Lake Valencia (Venezuela). *Arch. Hydrobiol.* **82:**347–358.

Jassby, A. D., and C. R. Goldman. 1974. Loss rates from a lake phytoplankton community. *Limnol. Oceanogr.* **19:**618–627.

Juday, C. 1921. Observations on the larva of *Corethra punctipennis* SAY. *Biol. Bull.* **40:**278–286.

Kalff, J., H. J. Kling, S. H. Holmgren, and H. E. Welch. 1975. Phytoplankton, phytoplankton growth and biomass cycles in an unpolluted and in a polluted polar lake. *Verh. Int. Ver. Limnol.* **19:**487–495.

Keen, R. 1973. A probabilistic approach to the dynamics of natural populations of Chydoridae (Cladocera, crustacea). *Ecology* **54:**520–534.

Kerfoot, W. C. 1974. Egg-size cycle of a cladoceran. *Ecology* **55:**1259–1270.

Kerfoot, W. C. 1975. The divergence of adjacent populations. *Ecology* **56:**1298–1313.

Kiefer, F. 1929. *Das Tierreich: Crustacea Copepoda II. Cyclopoida Gnathestoma.* Walter de Gruyter, Berlin. 102 pp.

Kiefer, F. 1938. Die von der Wallacea-Expedition gesammelten Arten der Gattung *Thermocyclops* Kiefer. *Int. Rev. Ges. Hydrobiol.* **38:**54–93.

Kosswig, C., and W. Villwock. 1964. Das problem der intralakustrischen Speziation im Titicaca—und in Lanaosee. *Verh. Dtsch Zool. Gesell.* **28:**95–102.

Lam, R. K., and B. Frost. 1976. Model of copepod filtering response to changes in size and concentration of food. *Limnol. Oceanogr.* **21:**490–500.

Lane, P. A. 1975. The dynamics of aquatic ecosystems: A comparative study of the structure of four zooplankton communities. *Ecol. Monogr.* **45:**307–336.

Lewis, W. M. Jr. 1973a. The thermal regime of Lake Lanao (Philippines) and its theoretical implications for tropical lakes. *Limnol. Oceanogr.* **18:**200–217.

Lewis, W. M., Jr. 1973b. A limnological survey of Lake Mainit, Philippines. *Int. Rev. Ges. Hydrobiol.* **58:**801–818.

Lewis, W. M., Jr. 1974. Primary production in the plankton community of a tropical lake. *Ecol. Monogr.* **44:**377–409.

Lewis, W. M., Jr. 1975. Distribution and feeding habits of a tropical *Chaoborus* population. *Verh. Int. Verein. Limnol.* **19:**3106–3119.

Lewis, W. M., Jr. 1976. Surface:volume ratio: implications for phytoplankton morphology. *Science* **192**:885–887.

Lewis, W. M., Jr. 1977a. Feeding selectivity of a tropical *Chaoborus* population. *Freshwater Biol.* **7**:311–325.

Lewis, W. M., Jr. 1977b. Net growth rate through time as an indicator of similarity among phytoplankton species. *Ecology* **58**:149–157.

Lewis, W. M., Jr. 1977c. Ecological significance of the shapes of abundance-frequency distributions for coexisting phytoplankton species. *Ecology* **58**:850–859.

Lewis, W. M., Jr. 1978a. A compositional, phytogeographical and elementary structural analysis of the phytoplankton in a tropical lake. *J. Ecol.* **66**:213–226.

Lewis, W. M., Jr. 1978b. Dynamics and succession of the phytoplankton in a tropical lake. *J. Ecol.* **66**:849–880.

Lewis, W. M., Jr. 1978c. Comparison of spatial and temporal variation in the zooplankton of a lake by means of variance components. *Ecology* **59**:666–671.

Lewis, W. M., Jr. 1978d. Analysis of succession in a tropical plankton community and a new measure of succession rate. Amer. *Natural.* **112**(984):401–414.

Lewis, W. M., Jr. 1978e. Spatial distribution of the phytoplankton in a tropical lake. *Int. Rev. Ges. Hydrobiol.* **63**:619–635.

Lewis, W. M., Jr. 1979. Evidence for stable zooplankton community structure gradients maintained by predation. In C. Kerfoot (ed.), *The Evolution and Ecology of Zooplankton Communities. Limnol. Oceanogr.* Special Symp. No. 3 (in press).

Likens, G. E., and J. J. Gilbert. 1970. Notes on the quantitative sampling of natural populations of planktonic rotifers. *Limnol. Oceanogr.* **15**:816–820.

Lindeman, R. L. 1942. The trophic dynamic aspect of ecology. *Ecology* **23**:399–418.

Lynch, M. 1977. Fitness and optimal body size in zooplankton populations. *Ecology* **58**:763–774.

Macan, T. T. 1977. The influence of predation on the composition of fresh-water animal communities. *Biol. Rev.* **52**:45–70.

MacDonald, W. W. 1956. Observations on the biology of chaoborids and chironomids in Lake Victoria and on the feeding habits of the "Elephant-snoutfish" (*Mormyrus kannume* Forsk) *J. Anim. Ecol.* **25**:36–53.

Makarewicz, J. C. 1974. The community of zooplankton and its production in Mirror Lake. Ph.D. Thesis, Cornell University.

Margalef, R. 1958. Temporal succession and spatial heterogeneity in phytoplankton. *In* A. Buzzati-Traverso (ed.), *Perspectives in Marine Biology,* pp. 323–347. University of California Press, Berkeley, California.

Margalef, R. 1967. Some concepts relative to the organization of plankton. *Oceanogr. Mar. Biol. Annu. Rev.* **5**:227–289.

Maynard Smith, J. 1978. *The Evolution of Sex.* Cambridge University Press. 222 pp.

McGowan, L. M. 1974. Ecological studies on *Chaoborus* (Diptera, Chaoboridae) in Lake George, Uganda. *Freshwat. Biol.* **4**:483–505.

McQueen, J. D. 1970. Grazing rates and food selection in *Diaptomus oregonensis* (Copepoda) from Marion Lake, British Columbia. *J. Fish. Res. Bd. Can.* **27**:13–20.

Melack, J. M. 1976. Primary productivity and fish yields in tropical lakes. *Trans. Am. Fish. Soc.* **105**:575–580.

Moriarty, D. J. W., J. P. E. C. Darlington, E. G. Dunn, C. M. Moriarty, and M. P. Tevlin. 1973. Feeding and grazing in Lake George, Uganda. *Proc. R. Soc. London B* **184**:299–319.

Myers, G. S. 1960. The endemic fish fauna of Lake Lanao, and the evolution of higher taxonomic categories. *Evolution* **14**:323–333.

Naumann, E. 1923. Spezielle Untersuchungen uber die Ernährungsbiologie des tierischen Limnoplanktons. II. Über den Nahrungserwerb und die naturlische Nahrung der Copepoden und die Rotiferen des Limnoplanktons. *Lunds Univ. Arsskr. n. f.* **19**:3–17.

Nauwerck, A. 1963. Die Beziehungen zwischen Zooplankton und Phytoplankton im See Erken. *Symb. Bot. Upsalienses* **17**:1–163.

Neill, W. E. 1975. Experimental studies of microcrustacean competition, community composition and efficiency of resource utilization. *Ecology* **56**:809–826.

O'Brien, W. J., N. A. Slade, and G. L. Vinyard. 1976. Apparent size as the determinant of prey selection by bluegill sunfish (*Lepomis machrochirus*). *Ecology* **57**:1304–1310.

Paffenhöffer, G. A. 1971. Grazing and ingestion rates of nauplii copepodids and adults of the marine planktonic copepod *Calanus helgolandicus*. *Mar. Biol.* **11**:286–298.

Paffenhöffer, G. A., and P. P. Harris. 1976. Feeding, growth and reproduction of the marine planktonic copepod *Pseudocalanus elongatus* Boeck. *J. Mar. Biol. Assn. U. K.* **56**:327–344.

Paloheimo, J. E. 1974. Calculation of instantaneous birthrate. *Limnol. Oceanogr.* **19**:692–694.

Parma, S. 1971. *Chaoborus flavicans* (Meigen) (Diptera, Chaoboridae): An autecological study. Ph.D. Dissertation, University of Gronigen, Rotterdam, Bronder-Offset n.v. 128 pp.

Parsons, T. R., M. Takahashi, and B. Hargrave. 1977. *Biological Oceanographic Processes*, 2nd ed. Pergamon Press, New York.

Pastorok, R. A. 1978. Predation by *Chaoborus* larvae and its impact on the zooplankton community. Doctoral Dissertation, University of Washington.

Patalas, K. 1971. Crustacean plankton communities in 45 lakes in the Experimental Lakes Area, Northwestern Ontario, *J. Fish. Res. Bd. Can.* **28**:231–244.

Patalas, K. 1975. The crustacean plankton communities of fourteen North American great lakes. *Verh. Int. Verein. Limnol.* **19**:504–511.

Pennak, R. 1957. Species composition of limnetic zooplankton communities. *Limnol. Oceanogr.* **2**:222–232.

Pianka, E. R. 1978. *Evolutionary Ecology*, 2nd ed. Harper and Row, New York.

Platt, T., and K. L. Denman. 1975. Spectral analysis in ecology. *Annu. Rev. Ecol. Syst.* **6**:189–210.

Platt, T., and C. Filion. 1973. Spatial variability of the productivity: biomass ratio for phytoplankton in a small marine basin. *Limnol. Oceanogr.* **18**:743–749.

Platt, T., L. M. Dickies, and R. W. Trites. 1970. Spatial heterogeneity of phytoplankton in a near-shore environment. *J. Fish. Res. Bd. Can.* **27**:1453–1473.

Porter, K. G. 1973. Selective grazing and differential digestion of algae by zooplankton. *Nature (London)* **244**:179–180.

Porter, K. G. 1975. Viable gut passage of gelatinous green algae ingested by *Daphnia*. *Verh. Int. Ver. Limnol.* **19**:2840–2850.

Pourriot, R. 1963. Utilisation des algue brunes unicellularies, pour l'élèvage des rotifères. *C. R. Hebd. Séanc. Acad. Sci. Paris* **256**:1603–1605.

Pourriot, R. 1965. Recherches sur l'ecologie des rotiferes. *Vie Milieu Suppl.* **21**:1–224.

Powell, T. M., P. J. Richerson, T. M. Dillon, G. A. Agee, B. J. Dozier, D. A. Godden, and L. O. Myrup. 1975. Spatial scales of current speed and phytoplankton biomass fluctuations in Lake Tahoe. *Science* **389**:1088–1090.

Prepas, E. and F. H. Rigler. 1978. The enigma of *Daphnia* death rates. *Limnol. Oceanogr.* **23**(5):970–988.

Ravera, O. 1954. La struttura demografica dei Copepodi del Lago Maggiore. *Mem. Ist. Ital. Idrobiol.* **8**:109–150.

Rey, J., and J. Capblancq. 1975. Dynamique des populations et production du zooplancton du lac de Port-Bielh (Pyrenees Centrales). *Ann. Limnol.* **11**:1–45.

Reynolds, C. S. 1976. Succession and vertical distribution of phytoplankton in response to thermal stratification in a lowland mere, with special reference to nutrient availability. *J. Ecol.* **64**:529–551.

Richerson, P. J., Widmer, C., Kittel, T. and Landa C. 1975. A survey of the physical and chemical limnology of Lake Titicaca. *Verh. Internat. Verein. Limnol.* **19**:1498–1503.

Rigler, F. H., and J. M. Cooley. 1974. The use of field data to derive population statistics of multivoltine copepods. *Limnol. Oceanogr.* **19**:636–655.

Rigler, F. H., M. E. MacCallum, and J. C. Roff. 1974. Production of zooplankton in Char Lake. *J. Fish. Res. Bd. Can.* **31**:637–646.

Ruttner, F. 1937. Limnologische Studien in einigen Seen der Östalpen. *Arch. Hydrobiol.* **32**:167–319.

Ruttner, F. 1952. Planktonstudien der Deutschen limnologischen Sunda-Expedition. *Arch. Hydrobiol. Suppl.* **21**:1–274.

Ruttner-Kolisko, A. 1974. Plankton Rotifers. Biology and Taxonomy. *Die Binnengewasser. Suppl.* **26**:1–146.

Schindler, D. W. 1969. Two useful devices for vertical plankton and water sampling. *J. Fish. Res. Bd. Can.* **26**:1948–1955.

Schindler, D. W. 1972. Production of phytoplankton and zooplankton in Canadian shield lakes. In Z. Kajak and A. Hilbricht-Ilkowska (eds.), *Productivity Problems of Freshwaters*, pp. 311–331. Proceedings of IBP–UNESCO Symposium, Kazimierz Dolny, Poland.

Slobodkin, L. B. 1961. *Growth and Regulation of Animal Populations.* Holt, Rinehart and Winston, New York. 184 pp.

Slobodkin, L. B. 1974. Prudent predation does not require group selection. *Amer. Nat.* **108**:665–678.

Sokal, R. R., and F. J. Rohlf. 1969. *Biometry.* W. H. Freeman, San Francisco. 776 pp.

Starkweather, P. L., and J. J. Gilbert. 1977. Feeding in the rotifer *Brachionus calyciflorus*. II. The effect of food density on feeding rates using *Euglena gracilis* and *Rhodotorula glutinis*. *Oecologia* **48**:133–139.

Stross, R. G., J. C. Neess, and A. D. Hasler. 1961. Turnover time and production of planktonic crustaceans in limed and reference portion of a bog lake. *Ecology* **42**:237–244.

Swift, M. C. 1976. Energetics of vertical migration in *Chaoborus trivittatus* larvae. *Ecology* **57**(5):900–915.

Wiebe, P. H. 1970. Small-scale distribution in oceanic zooplankton. *Limnol. Oceanogr.* **15**:205–217.

Williams, G. C. 1975. *Sex and Evolution.* Princeton University Press, Princeton. 200 pp.

Wilson, D. S. 1973. Food size selection among copepods. *Ecology* **54**:909–914.

Winberg, G. G. 1971. *Methods for the Estimation of Production of Aquatic Animals.* Academic Press, New York. 175 pp.

Winkler, R. L., and W. L. Hays. 1975. *Statistics, Probability, Information, and Decision.* Holt, Rinehart & Winston, New York. 475 pp.

Woltereck, R. 1941. Die Seen und Inseln der "Wallacea" Zwischenregion und ihre Endemische Tierwelt. *Int. Rev. Ges. Hydrobiol.* **41**:1–203.

Wright, J. C. 1958. The limnology of Canyon Ferry Reservoir I. Phytoplankton–zooplankton relationships in the euphotic zone during September and October, 1956. *Limnol. Oceanogr.* **3**:150–159.

Zaret, T. M. 1972. Predator–prey interaction in a tropical lacustrine ecosystem. *Ecology* **53**:248–257.

Zaret, T. M. 1975. Strategies for existence of zooplankton prey in homogeneous environments. *Verh. Int. Ver. Limnol.* **19**:1484–1489.

Zaret, T. M., and J. S. Suffern. 1976. Vertical migration in zooplankton as a predator avoidance mechanism. *Limnol. Oceanogr.* **21**:804–813.

Zeuthen, E. 1953. O_2 uptake as related to size of organisms. *Quart. Rev. Biol.* **28**:1–12.

Author Index

Subject Index